‘Mechanical Handling with Precisio

(Amendments to pages 14 and 21)

Please note: it is necessary to make the following amend figs 2.2 and 2.11. These are indicated below in bold type appropriate portion and amend your copy.

Fig 2.2 Final chain selection formulae

Page 14

Material sliding

Horizontal, to 8°

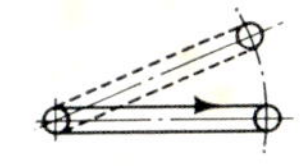

Final selection Chain Pull (lbf) =

$$\mathbf{Wf_4} + \frac{W_A}{2}(f_2+f_3)$$

Over 8°, to vertical

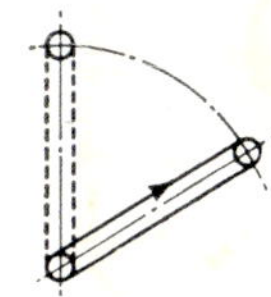

Final selection Chain Pull (lbf) =

$$\mathbf{Wf_4} + \frac{W_A}{2}(f_2+f_3)$$

Material carried

Horizontal, to 8°

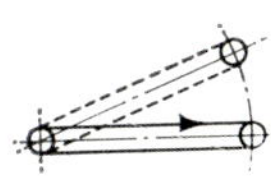

Final selection Chain Pull (lbf) =

$$Wf_2 + \frac{W_A}{2}(f_2+f_3)$$

Over 8°, to vertical

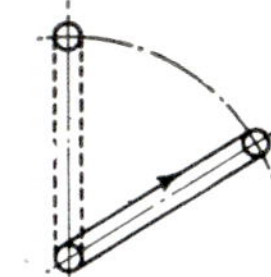

Final selection Chain Pull (lbf) =

$$\mathbf{Wf_2} + \frac{W_A}{2}(f_2+f_3)$$

Fig 2.11 Horsepower formulae

Page 21

Material sliding

Horizontal, to 8°

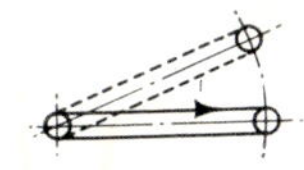

Horsepower

$$hp = \frac{S \times (Wf_4 + W_A\mathbf{f_5})}{33{,}000}$$

Over 8°, to vertical

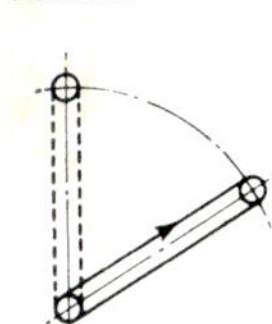

Horsepower

$$hp = \frac{S \times (Wf_4 + W_Af_5)}{33{,}000}$$

Material carried

Horizontal, to 8°

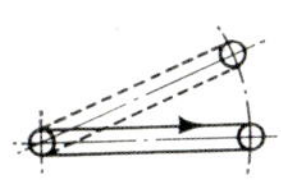

Horsepower

$$hp = \frac{S \times (Wf_2 + \frac{W_A}{2}\mathbf{f_5})}{33{,}000}$$

Over 8°, to vertical

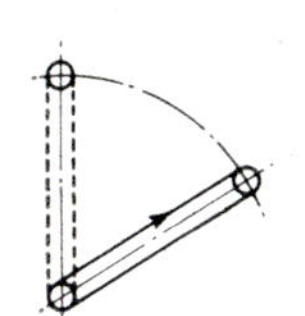

Horsepower

$$hp = \frac{S \times (Wf_2 + \frac{W_A}{2}f_5)}{33{,}000}$$

Where:—

W = Weight of material on conveyor (lb)

W_A = Actual total weight of chains and conveying attachments **(lb)**

S = Chain speed (ft/min)

f_2 = See Table 2.6

f_3 = See Table 2.7

f_4 = See Table 2.2

f_5 = See Table 2.9

MECHANICAL HANDLING
WITH PRECISION CONVEYOR CHAIN

Mechanical Handling with Precision Conveyor Chain

Edited by
L. Jones C.Eng., M.I.Mech.E.
Chief Sales Engineer, Renold Limited

Foreword by
J. W. Leech
President of the Mechanical Handling Engineers Association

HUTCHINSON OF LONDON

HUTCHINSON & CO (*Publishers*) LTD
3 Fitzroy Square, London W1

London Melbourne Sydney Auckland
Wellington Johannesburg Cape Town
and agencies throughout the world

First published 1971

This book has been set in Times type, printed in Great Britain on kirbycote paper by Anchor Press, and bound by Wm. Brendon, both of Tiptree, Essex

ISBN 0 09 106990 4

Contents

All units of measurement and weight are expressed in INCH/POUND units, but conversion to metric units is simplified by the provision of a comprehensive set of conversion tables.

Foreword

by

J. W. Leech

President of the Mechanical Handling Engineers Association

The introduction of mechanical handling into industry might be termed the Second Industrial Revolution. In this, the use of chain conveyors and elevators has played an ever increasing part.

In the latter part of the nineteenth century and up to the early part of the twentieth century, cast link chains were the only ones available for this purpose.

Hans Renold Limited saw the immense possibilities of the use of steel precision conveyor chains in the development of industrial application and in the late 1920's they produced their first complete range. To enable mechanical handling engineers to make full use of these, they also produced their first catalogue including technical guidance on selection and usage.

It is of particular interest to me as the one responsible for sales promotion of the chains in these early years to note that in this excellent treatise which they have now produced, much of the theory and practice that was recommended in those early days has been proved a sound basis for present-day practice as set out herein.

The present work covers a much wider field of application and gives precise practical data based on experience gained over the past 40 years which I know will prove invaluable to handling engineers and students of today and the future.

It is simply and clearly set out, which will enable quick reference to be made and which will also help systems engineers in preparing projects for the consideration of factory managements.

Renold Limited are in a unique position to service the mechanisation of industries as they make all the transmission products which, together with their precision conveyor chains, make up the essentials for a wide range of mechanical handling applications and no industry interested in mechanical handling can afford to be without a copy of this excellent work on the subject.

Introduction

From the pyramids to the railway revolution, muscle-power of men and animals has moved goods and materials. But throughout history, machines however primitive have played some part, becoming more and more versatile.

Within the immediate past, mechanical handling has emerged as a manufacturing industry in its own right, of considerable size and with countless applications. This is a consequence of its coverage, which now ranges from the simplest store conveyor system to the largest flow line production layouts, and also includes the movement of personnel by lifts, escalators and platforms.

The basic objects of mechanical handling can be stated simply as the reduction or elimination of fatigue and drudgery; plus speed and economy in moving components, goods and articles, and bulk materials of all kinds. Added to these is the handling of heavy and bulky objects where mechanical aid is unavoidable, and the increase in industrial plant efficiency as a consequence of bringing components to the right place at the required time, without hold-up or distraction.

Amongst the most widely used types of handling equipment are conveyors, elevators and similar assemblies. These can take many forms, employing as their basic moving medium both metallic and non-metallic components or a mixture of the two. This book deals solely and comprehensively with the application of precision conveyor chain to mechanical handling. The significance of this design of chain must be appreciated since other types and varieties of chain are successfully used.

For the great majority of applications the precision conveyor chain in its many variations, and when fitted with suitable attachments, provides a highly efficient propulsion and/or carrying medium, having many advantages over other types.

It must be emphasised that this book does not purport to answer all the problems associated with chain application to mechanical handling equipment. As new materials and requirements appear, fresh problems arise and development takes place to deal with them. It is to the advantage of all concerned to maintain the closest contact with chain manufacturers, whenever new factors call for advice or investigation.

Wherever possible the terms used to describe types of conveyors are in accordance with BS3810: Part 2: 1965 Glossary of Terms Used in Materials Handling.

Throughout the book Renold nomenclature and trade names have been used in the description of chains and associated drive components.

List of Symbols Used

a = Distance from pivot point to contact point on tray (in)
b_L = Distance from backing guide to centre of gravity of load (in)
b_T = Distance from backing guide to centre of gravity of tray (in)
bf_1 = Shroud width
bf_2 = Face to wheel centreline
bf_3 = Distance through boss
c = Minimum clearance (in)
d = Pitch circle diameter of wheel
da = Top diameter of wheel
db = Boss diameter
de = Bore diameter
dg = Shroud diameter of wheel
e = Naperian logarithm base (2·718)
f = Rate of acceleration (ft/sec^2)
f_1, f_2 etc = Friction factors
f_S = Maximum extreme fibre stress (lbf/in^2)
g = Acceleration due to gravity (32·2 ft/sec^2)
h = Height (ft)
h_A = Height of aperture (in)
h_C = Height from low to high level (ft)
h_M = Height of material (in)
h_T = Height from centre of chain to centre of trolley roller (in)
h_U = Height of pusher (in)
hp = Horsepower
i = Maximum slope (radians)
id = Inside diameter of roller (in)
k_W = Radius of gyration of wheel (in)
l_A = Length of chain in circuit (ft)
l_B = Length of bunker (ft)
l_C = Conveyor centres (ft)
l_I = Distance between idlers (ft)
l_S = Length of various sections of conveyor (ft)
l_T = Spacing of trucks (ft)
m = Mass
od = Outside diameter of roller (in)
p = Chain pitch (in)
p_R = Pitch of trolley rollers (in)
p_S = Spacing of bucket (in)

List of Symbols Used

r = Minimum radius (in)
r_B = Radius of bend (ft)
r_M = Radius of centre of gravity of material (ft)
r_W = Radius of wheel (in)
ton/h = Tons per hour
t = Time lapse after material leaves bucket (sec)
t_{max} = Time to attain max speed (sec)
v = Chain speed (ft/sec)
v_M = Speed of material at centre of gravity of bucket (ft/sec)
v_{max} = Maximum speed (ft/sec)
w = Weight of chain(s) and conveying attachments (lb/ft)
w_C = Weight of material, chain and conveying attachments (lb/ft)
w_M = Weight of material on conveyor (lb/ft)
w_S = Weight of slats and fixing bolts (lb/ft)
y = Distance from neutral axis to extreme fibre (in)

C = Elevator capacity (lb/min)
E = Modulus of Elasticity
F = Force (lbf)
F_A = Accelerating force (lbf)
F_C = Centrifugal force (lbf)
F_F = Force to overcome friction (lbf)
F_M = Factor depending on the material handled
F_R = Resultant force (lbf)
F_T = Total force (lbf)
I = Moment of inertia (in^4)
J = Chain sag (ft)
K = Constant
K_W = Constant for bogie wheel
L = Roller load (lbf)
L_B = Length of beam (in)
L_C = Total length of chain (ft)
M = Bending moment (lbf in)
N_S = Number of teeth in snub wheel
P = Chain pull (lbf)
P_R = Maximum permissible pressure between chain roller and bush (lbf/in^2)
S = Chain speed (ft/min)
T_A = Actual tension load (lbf)
T_T = Theoretical tension load (lbf)
V = Rubbing speed between chain roller and bush (ft/min)
W = Weight of material on conveyor (lb)

W_A	=	Actual weight of chains and conveying attachments (lb)
W_B	=	Weight of material in bucket (lb)
W_C	=	Weight of tray (lb)
W_D	=	Weight of material on tray (lb)
W_E	=	Estimated weight of chains and conveying attachments (lb)
W_G	=	Weight of bogie (lb)
W_M	=	Weight of material supported by conveyor (lb)
W_S	=	Weight of material on staybar (lbf)
W_T	=	Weight of truck (lb)
W_W	=	Weight of each axle and its wheels (lb)
β	=	Angle of discharge (degrees)
θ	=	Bend angle (radians)
μ	=	Coefficient of friction between chain and track
μ_R	=	Overall coefficient of friction of chain rolling on track
μ_{R1}	=	Coefficient of rolling friction of chain roller on track
μ_{R2}	=	Overall coefficient of friction of support rollers running on track
μ_{R3}	=	Coefficient of rolling friction between chain roller and load
μ_{R4}	=	Overall coefficient of trolley roller on track
μ_{R5}	=	Overall coefficient of truck roller on track
μ_{R6}	=	Overall coefficient of friction of bogie wheel on track
μ_S	=	Coefficient of friction of chain sliding on track
μ_{S1}	=	Coefficient of sliding friction between body and plane
μ_{S2}	=	Coefficient of sliding friction between material and conveyor
μ_{S3}	=	Coefficient of friction of material sliding against itself
μ_{S4}	=	Coefficient of sliding friction between chain bush and roller
μ_{S5}	=	Coefficient of sliding friction between tray and guide
μ_{S6}	=	Coefficient of sliding friction between chain and material
ϕ	=	Conveyor incline (degrees)

1

The Precision Conveyor Chain

Design and range

1.1

Roller chain has been employed as an efficient means of transmitting power since it was invented by Hans Renold in 1880. Later the principle was applied to conveyor chain giving the same advantages of precision, heat-treated components to resist wear, high strength to weight ratio and high mechanical efficiency.

A precision conveyor chain is made up of a series of inner and outer links. Each link comprises components manufactured from a material best suited to its function in the chain; the various parts are shown in Fig 1.1. An inner link consists of a pair of inner plates which are pressed on to cylindrical bushes, whilst on each bush a free fitting roller is normally assembled. Each outer link has a pair of outer plates which are pressed on to bearing pins and the ends of the pins are then riveted over the plate.

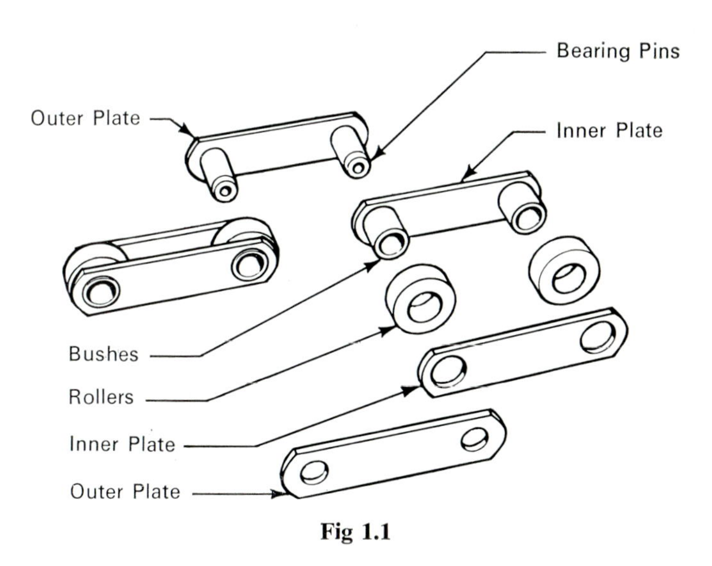

Fig 1.1

From the foregoing it will be seen that a length of chain is a series of plain journal bearings free to articulate in one plane. When a chain articulates under load the friction between pin and bush, whilst inherently low because of the accurate finish on the components, will tend to turn the bush in the inner plates and similarly the bearing pin in the outer plate. To prevent this the bush and pin are force fitted into the chain plates. Close limits of accuracy are applied to the diameter of plate holes, bushes and bearing pins resulting in high torsional security and rigidity of the mating components. Similar standards of accuracy apply to the pitch of the holes in the chain plates.

To ensure optimum wear life the pin and bush are hardened. The bush outside diameter is hardened to contend with the load carrying pressure and gearing action both of which are imparted by the chain rollers. Chain roller material and diameter can be varied and are selected to suit applicational conditions; guidance in roller selection is given in Chapter 2. Materials used in chain manufacture conform to a closely controlled specification. Manufacture of components is similarly controlled both dimensionally and with regard to heat treatment.

Basically conveyor chains are specified by breaking load. For each breaking load a range of chain pitches is available, the minimum pitch being governed by the need for adequate tooth strength. Maximum pitch is normally determined by general chain rigidity, although this may be increased by incorporating a strengthening bush between the link plates. Wherever possible it is desirable to use catalogued stock chain pitches as these afford advantages in price and delivery. Dimensions and data for a typical range of precision conveyor chains are given in Tables 1.1 and 1.2.

To appreciate the versatility of precision conveyor chain it is necessary to examine the various types of conveying equipment. This is done in later chapters, but at this stage it is useful to illustrate the available variations in basic chain assembly and give brief descriptions of the usual types of conveyor attachments.

Two main patterns of chain are available—hollow bearing pin and solid bearing pin. The hollow type affords the ready facility of fitting attachments through the hollow pin. If attachments are bolted rigidly through two adjacent pins then this must be done on the outer link assembly, otherwise correct articulation of the chain will be prevented.

1.2 Chain attachments

An attachment is any part fitted to the basic chain to adapt it for a particular conveying duty. Attachments may be an integral part of the chain plate or they can be built into the chain as a replacement of the normal link. Other attachments are either welded, riveted or bolted to the chain plates.

Some types of attachments are designated by a letter of the alphabet. The origin of this identification is obscure and is not helped by the fact that different manufacturers' nomenclature do not always agree. In describing the various attachments available the Renold nomenclature has been used.

K ATTACHMENTS

These are the most popular type of attachment, being used on slat and apron conveyors, also bucket elevators etc. As shown in Fig 1.2 they provide a platform parallel to the chain and bearing pin axes. They are used for securing slats and buckets etc. to the chain. Either 1 or 2 holes are normally provided in the platform, being designated K1 or K2 respectively. K attachments can be incorporated on one or both sides of the chain. For the more important stock pitches where large quantities justify the use of special manufacturing equipment the attachments are produced as an integral part of the chain, as shown in Fig 1.2(a). Here the platform is an extension of the chain plate itself and then bent over.

On higher strength chains or where only small quantities are involved, separate attachments are used, as shown in Fig 1.2(b). These are either welded, riveted or bolted to the chain depending on the particular chain series and the application. Alternatively, (see Fig 1.2(c)), K attachments may be bolted through the hollow bearing pin or by using the special outer links with extended and screwed bearing pin ends.

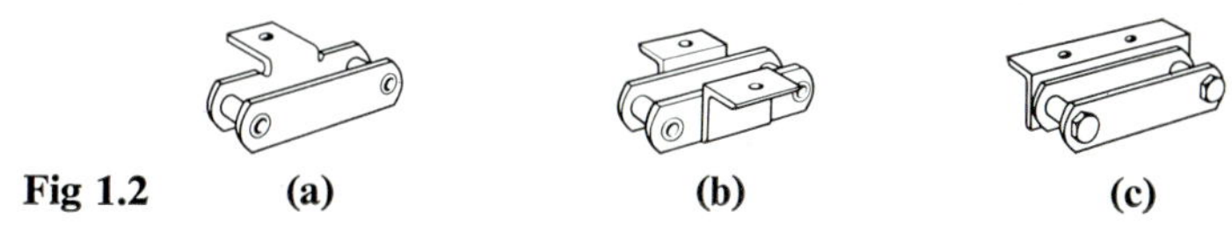

Fig 1.2 **(a)** **(b)** **(c)**

(a) K1 bent over attachment **(b)** K1 attachment, welded, riveted or bolted to link plate **(c)** K2 attachment bolted through hollow bearing pin

F ATTACHMENTS

These attachments as shown in Fig 1.3 are frequently used for pusher and scraper applications. They comprise a wing with a vertical surface at right angles to the chain. They can be fitted to one or both sides of the chain and are secured by either welding, riveting or bolting. Each wing can be provided with one or two holes, being designated F1 or F2 respectively.

Fig 1.3 **(a)** **(b)**

(a) F1 attachment, welded, riveted, or bolted to link plate
(b) F2 attachment, welded, riveted, or bolted to link plate

SPIGOT PINS AND EXTENDED BEARING PINS

Both types are used on pusher and festoon conveyors and tray elevators etc. Spigot pins may be assembled through hollow bearing pins, inner and outer links. When assembled through link plates a spacing bush is necessary to ensure that the inside width of the chain is not reduced. Gapping of the wheel teeth is necessary to clear the bush.

Solid bearing pin chains can have a similar extension at the pitch point by incorporating extended pins. Both spigot pins and extended pins, as shown in Fig 1.4, can be case-hardened on their working diameter.

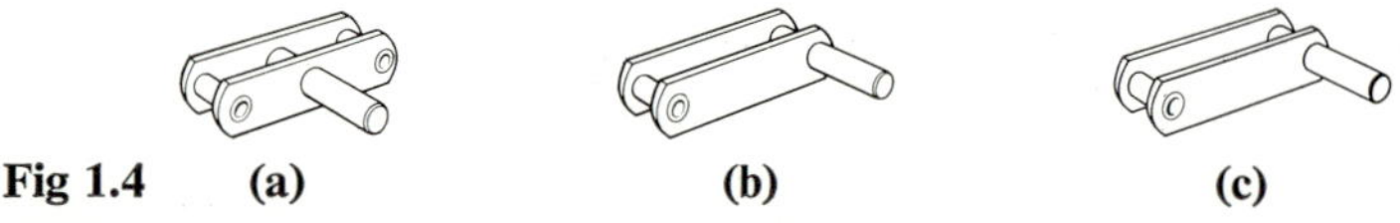

Fig 1.4 **(a)** **(b)** **(c)**

(a) Spigot pin assembled through outer or inner link **(b)** Spigot pin bolted through hollow bearing pin **(c)** Extended bearing pin

STAYBARS

Types of mechanical handling equipment that use staybars are pusher, wire mesh and festoon conveyors, the staybars being assembled in the same manner as spigot pins. When assembled through link plates a spacing bush and gapping of the wheel teeth are necessary.

The plain bar-and-tube type shown in Fig 1.5 has the advantage that the staybar can be assembled with the chains in situ by simply threading the bar through the chains and tube. The shouldered bar type has a greater carrying capacity than the bar-and-tube type. Staybars are normally used for either increasing overall rigidity by tying two chains together, maintaining transverse spacing of the chains or supporting a load.

Fig 1.5 **(a)** **(b)**

(a) Staybars bolted through hollow bearing pin **(b)** Staybars assembled through outer or inner link

G ATTACHMENTS

As shown in Fig 1.6 this attachment takes the form of a flat surface positioned against the side of the chain plate and parallel to the chain line. It is normally used for bucket elevators and pallet conveyors. When the attachment is integral with the outer plate then the shroud diameter on the chain wheel will have to be removed to clear the plate. G attachments are normally only fitted to one side of the chain.

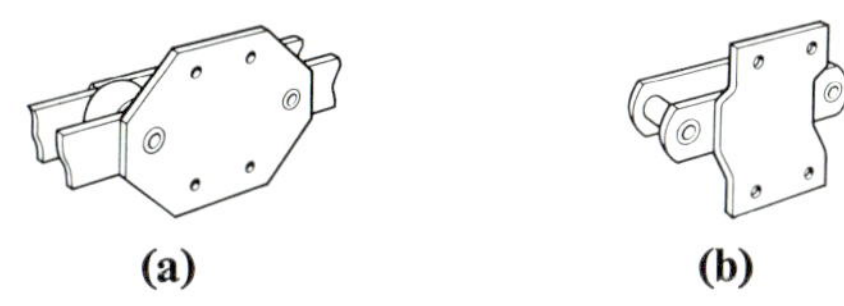

Fig 1.6 **(a)** **(b)**

(a) G2 attachment outer plate **(b)** G2 attachment, welded or riveted to link plate

L ATTACHMENTS

These have some affinity with the F type; being in a similar position on the chain, they are however confined to lower strength chains. A familiar application is the box scraper conveyor. As shown in Fig 1.7 the attachment is integral with the outer plate, being extended beyond one bearing pin hole and then bent over. The attachment can be plain or drilled with one or two holes, being designated L0, L1 or L2 respectively; they can be supplied on one or both sides of the chain. With this type of attachment the chain rollers are normally equal to the plate depth.

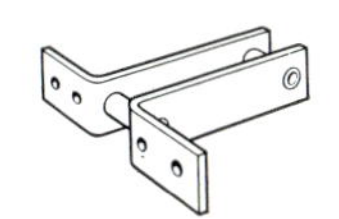

Fig 1.7 L2 attachment

S AND PUSHER ATTACHMENTS

These are normally used on dog pusher conveyors. As shown in Fig 1.8 the S attachment consists of a triangular plate integral with the chain outer plates; they can be assembled on one or both sides of the chain. They may also be assembled at the inner link position. S attachments are intended for light duty; for heavier duty a pair of attachments on one link is connected by a spacer block to form a pusher attachment. This increases chain rigidity and pushing area.

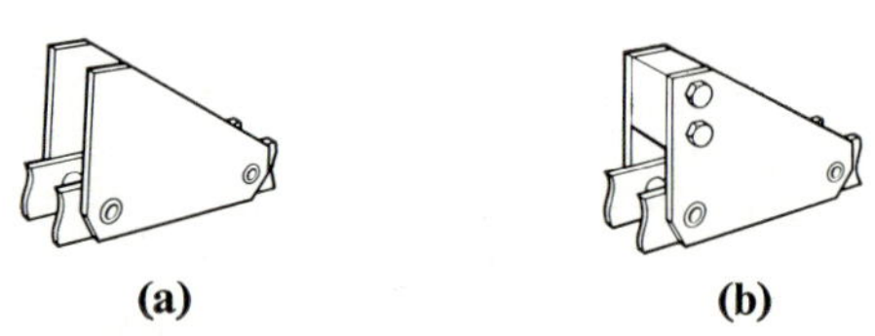

Fig 1.8 **(a)** **(b)**

(a) S attachment outer plate: assembled on one or both sides of chain as required **(b)** Pusher attachment

MALLEABLE SCRAPERS

These have been specifically designed for box scraper conveyors. They differ from the basic link design in that they comprise a casting in the form of a cross. As shown in Fig 1.9 one arm of the cross replaces the inner link and is drilled to receive the bearing pins of the adjacent outer link. The arm at 90° to the chain has inclined faces which impart a ploughing action and induce material flow. Chain wheel rims must be gapped to clear the scraper attachment.

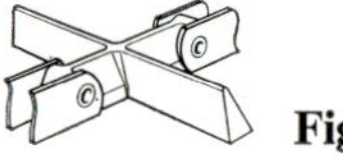

Fig 1.9 Malleable scraper

DEEP LINK PLATES

These have a greater plate depth than the basic chain, thus providing a continuous carrying edge above the chain rollers, as shown in Fig 1.10. The plates are normally arranged as a continuous assembly but deep plates can be fitted at intervals into standard chain.

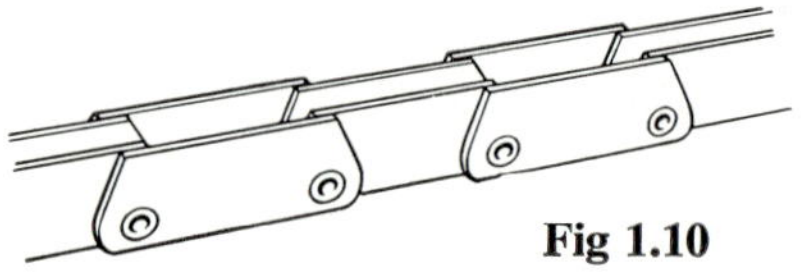

Fig 1.10

DRILLED LINK PLATES

Reference has already been made to attachments which are secured to link plates by bolting or riveting. Plates with single holes as shown in Fig 1.11(a) are associated with the fitting of staybars or spigot pins. Where G or K attachments are to be fitted then link plates with two holes as shown in Fig. 1.11(b) are used. When attachments are fitted to inner links then countersunk bolts or rivets must be used to provide wheel tooth clearance.

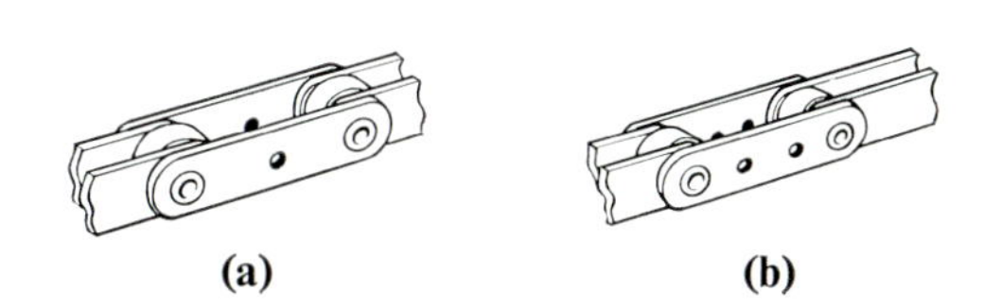

Fig 1.11 **(a)** **(b)**

BIPLANAR LINKS

As the name implies this type of link allows chain articulation in two planes, a requirement of many overhead conveyors. As shown in Fig 1.12, the links are assembled between normal outer links. At each end of the biplanar link is a solid pin, encircled by a U plate. The ends of the U are holed for a bush which is press fitted in the normal manner. Various types are available but they only differ in the method of plate retention.

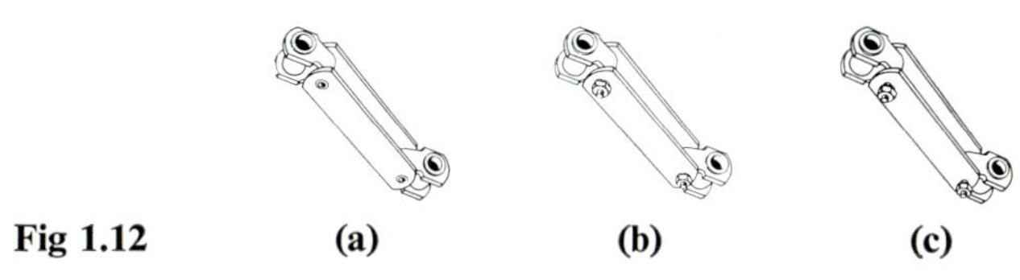

Fig 1.12 **(a)** **(b)** **(c)**

(a) Riveted construction **(b)** One plate detachable **(c)** Extended bearing pins one or both sides for attachments

OUTBOARD ROLLERS

The main reasons for using outboard rollers are that they increase roller loading capacity of the chain and provide a stabilised form of load carrier. A typical application is the pusher conveyor. As shown in Fig. 1.13 the outboard rollers are fixed to the chain by bolts which pass through hollow bearing pins. Outboard rollers have the advantage that they are easily replaced in the event of wear and allow the chain rollers to be used for gearing purposes only.

Fig 1.13 Outboard rollers

1.3 Chain joints

Conveyor chain is normally supplied in convenient handling lengths, these being joined by means of an outer connecting link (No. 107 joint), see Fig 1.14. This is supplied with one plate detached, the pins on that side being an interference fit in the plate holes. The end of the pin is softened to facilitate hand riveting after the plate has been pressed onto the pin. The No. 107 link is available for solid and hollow pin chains. The detachable link for final connecting up or removal of the chain can be obtained in several forms. On the No. 69 joint one plate is loose fit, the other plate being an interference fit on the pins and secured by riveting. Types of joints are illustrated in Fig 1.14.

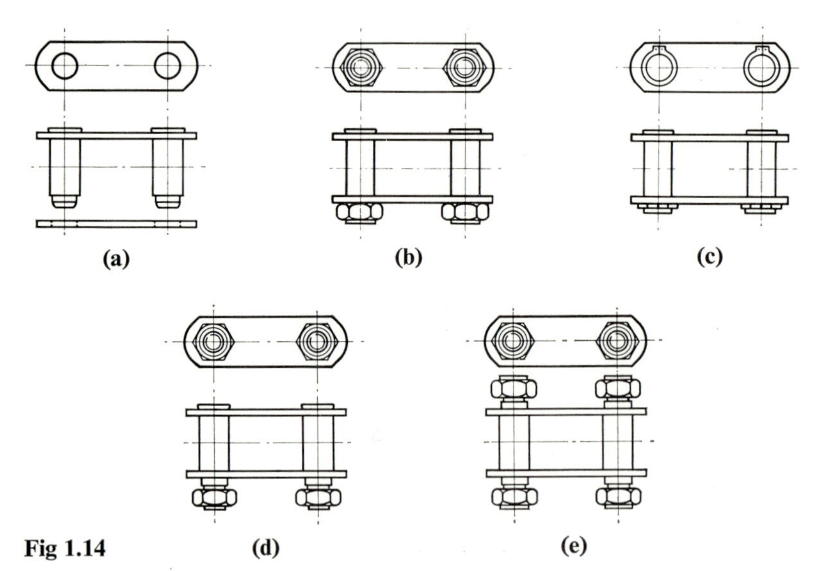

Fig 1.14

(a) No. 107 Riveted one side only **(b)** No. 69 Riveted one side: nuts one side **(c)** Riveted one side: circlips one side **(d)** Riveted one side: extended pins and nuts one side **(e)** Extended pins and nuts both sides

Pairing and matching of chains 1.4

When chains are required to run together they can be paired; this ensures that attachments on the chains are in line. If such chains are delivered in shorter lengths for ease of handling then these are tagged for identification on re-assembly. On some applications, such as escalators, it is necessary for the strands of chain to be matched to maintain a more precise relationship with each other; this is possible by accurate grading of each short length of chain by the manufacturer.

Advantages of precision conveyor chain 1.5

These can be summarised as follows:—
Large bearing areas and hardened components promote maximum life. Low friction is due to accurate finish of the components. Inclusion of chain roller and high strength to weight ratio gives lighter chain selection and lower power consumption. Use of high grade materials ensures reliability on onerous and arduous applications. The facility to obtain a variety of pitches with each chain breaking strength and a variation in attachment types provide adaptability. Accuracy of components provides consistency of operation, accurate gearing and low wheel tooth wear. The latter is particularly important in multi-strand systems where load distribution is important.

2

General Selection Procedure

2.1 Basic requirements

To enable the most suitable chain to be selected for a particular application it is necessary to know full applicational details such as the following:—

Type of conveyor.

Conveyor centre distance and inclination from the horizontal.

Type of chain attachment, spacing and method of fixing to the chain.

Number of chains and chain speed.

Details of conveying attachments, e.g. weight of slat.

Description of material, weight and size.

Method of feed and rate of delivery.

2.2 Selection of chain pitch

In general the largest stock pitch possible consistent with correct operation should be used for any application, since economic advantage results from the use of the reduced number of chain components per unit length. Other factors include size of bucket or slats etc., chain roller loading (see Section 2.4) and the necessity for an acceptable minimum number of teeth where space restriction exists.

2.3 Solid versus hollow bearing pin chain

Solid bearing pin chains are preferred on the following applications:—

Long conveyors where structural 'snaking' due to floor undulations etc. may bring the chains into rubbing contact with the guide tracks.

Applications where off-set or impact loading could tend to force the outer links off the bearing pin neck ends.

Corrosive applications where the hollow bearing pin would allow the ingress and retention of the corrosive agent.

Fairly high linear chain speed. The increased weight for the solid pin chain assists in maintaining correct tracking and promotes roller rotation as opposed to skidding.

Hollow pin chains are used in the following conditions:—

Where attachments utilise the hollow pin feature.

Clean and dry conditions.

Where reduced weight is required, while still maintaining adherence of the chain roller on the track.

Elevated temperature conditions, i.e. up to 300°C where the hollow pin affords sufficient surface area to promote natural cooling.

Chain selection by formula method 2.4

The two methods of selection are by formula and by section. The formula method being for use with simple conveyors and is based on the following assumptions:—

The load on a horizontal conveyor is carried on the upper or lower strand(s).

On vertical conveyors the load is being transported upwards.

The selection sequence is identical for both as follows:—

Preliminary calculation of load pull
Safety factor
Preliminary chain selection
Roller loading
Final selection

In a conveyor system the total load pull in the chain is derived from the combination of component pulls required to maintain motion. These being to move the load, to move the conveying attachments (e.g. slats etc.), to move the chains and to turn the wheels. Calculation of chain pull is carried out in two stages; a preliminary selection using an estimated chain weight and an approximate coefficient of friction. Later a final calculation can be made using the catalogued weight of chain and actual chain coefficient of friction.

PRELIMINARY CALCULATION OF CHAIN PULL

First estimate the weight of material (W) on the conveyor or elevator. Where the load is given in tons per hour (ton/h) then:—

$$W\ (\text{lb}) = \frac{\text{ton/h} \times 2{,}240 \times \text{Conveyor centres (ft)}}{60 \times \text{Conveyor speed (ft/min)}}$$

$$= \frac{\text{ton/h} \times \text{Conveyor centres (ft)} \times 37{\cdot}3}{\text{Conveyor speed (ft/min)}}$$

To this should be added any additional loads that might be significant, e.g. acceleration and load pick-up. For example on bucket elevators an additional dredging pull is caused by the buckets picking up material in the elevator boot (see Page 147).

The total weight of the conveying attachments (e.g. slats, buckets, etc.) should now be determined. At this stage the weight of the chain should be added, unfortunately this cannot be determined exactly until the chain strength is established. To overcome this the chain weight is assumed to be equal to the weight of the conveying attachments. A preliminary selection can now be made using the formulae given in Fig 2.1 and factors given in Tables 2.1 and 2.2.

Fig 2.1 Preliminary chain selection formulae

Material sliding

Horizontal, to 8°

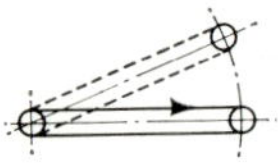

Preliminary selection Chain Pull (lbf)=

$Wf_4 + W_E f_l$

Over 8°, to vertical

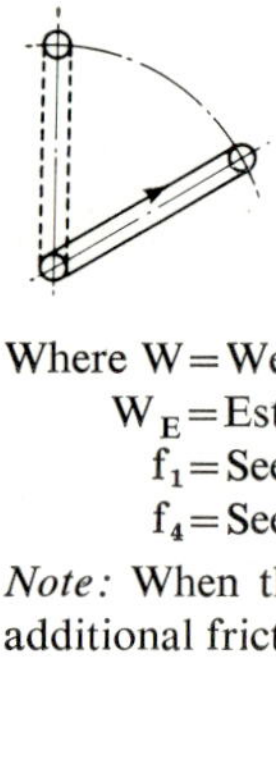

Preliminary selection Chain Pull (lbf)=

$Wf_4 + \frac{W_E}{2} f_l$

Material carried

Horizontal, to 8°

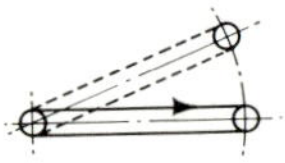

Preliminary selection Chain Pull (lbf)=

$Wf_l + W_E f_l$

Over 8°, to vertical

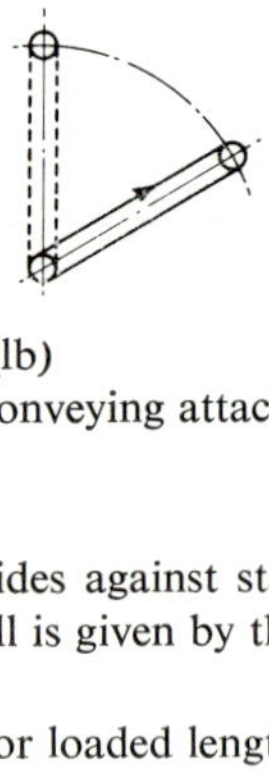

Preliminary selection Chain Pull (lbf)=

$Wf_l + \frac{W_E}{2} f_l$

Where W = Weight of material on conveyor (lb)

W_E = Estimated weight of chains and conveying attachments (lb)

f_1 = See Table 2.1

f_4 = See Table 2.2

Note: When the material being conveyed slides against stationary skirt plates an additional frictional pull is imposed. This pull is given by the following formula:—

$$\frac{h_M{}^2}{F_M} \text{ (lbf/ft) of conveyor loaded length}$$

Where h_M = Height of material (in)

F_M = Factor depending on material handled

Typical F_M values for material against steel plates are:—

Cement clinker 8, Clay (compact) 48, Coal (bituminous dry fines and lumps) 30, Gravel, stone 8, Iron ore (crushed) 4, Sugar cane 70, Wood chips, pulpwood 48.

SAFETY FACTOR

The next stage in chain selection is to determine a suitable safety factor which will provide an acceptable chain life by limiting the bearing pressure between pin and bush. Allowable bearing pressures are dependent upon the conditions under which the chain operates, this being generally dictated by the nature of material handled and the effectiveness of lubrication.

Each application requires individual consideration but for guidance safety factors are given in Table 2.3.

When a conveyor is started from rest fully loaded, the load pull in the chain will be greater than necessary to maintain motion. For the majority of applications this is catered for in the normal selection procedure. Start-up load can be up to 50% in excess of running load but this is usually infrequent compared with the running time of the installation. Steel roller chains can be subjected to a starting load of 20% of the chain strength.

PRELIMINARY CHAIN SELECTION

It is now possible to estimate the breaking strength of chain required although this will later need to be checked.

$$\text{Chain strength} = \frac{\text{Chain pull} \times \text{Safety factor}}{\text{Number of chains used}}$$

ROLLER LOADING

Under conditions of high unit loads it may be necessary to fit chains of higher strength than required by the tensile pull in order to provide adequate roller loading capacity.

Again the aim is to limit the resulting bearing pressure between the chain roller and bush to ensure satisfactory life (see Table 2.4).

The following give a general indication of maximum working loads applied to one roller.

V = Rubbing speed between chain roller and bush (ft/min)

$$= \frac{\text{Chain speed (ft/min)} \times \text{Roller bore (in)}}{\text{Roller diameter (in)}}$$

P_R = Maximum permissible pressure between chain roller and bush (lbf/in^2)

$$= \frac{\text{Roller load (lbf)}}{\text{Roller bore (in)} \times \text{Roller length (in)}}$$

To simplify roller loading calculations Table 2.5 gives actual roller loadings for the chains listed in Tables 1.1 and 1.2 and operating under average conditions.

If roller loading capacity of the chain provisionally selected is adequate then final chain selection can proceed. Should, however, roller loading be insufficient then three alternative solutions are available. A smaller pitch chain can be used, a chain with increased roller loading capacity can be employed or outboard rollers can be considered.

FINAL CHAIN SELECTION

Fig 2.2 Final chain selection formulae

Material sliding

Horizontal, to 8°

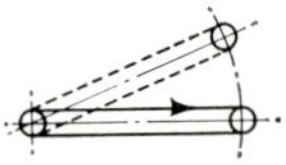

Final selection Chain Pull (lbf)=

$$Wf_2 + \frac{W_A}{2}(f_2 + f_3)$$

Over 8°, to vertical

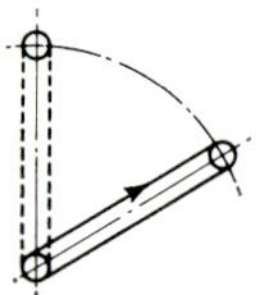

Final selection Chain Pull (lbf)=

$$Wf_2 + \frac{W_A}{2}(f_2 + f_3)$$

Material carried

Horizontal, to 8°

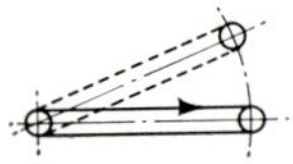

Final selection Chain Pull (lbf) =

$$Wf_2 + \frac{W_A}{2}(f_2 + f_3)$$

Over 8°, to vertical

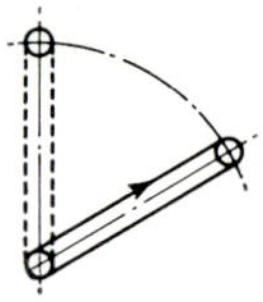

Final selection Chain Pull (lbf)=

$$Wf_4 + \frac{W_A}{2}(f_2 + f_3)$$

Where W=Weight of material on conveyor (lb)
W_A=Actual weight of chains and conveying attachments (lb)
f_2=See Table 2.6
f_3=See Table 2.7
f_4=See Table 2.2

Note: When the material being conveyed slides against stationary skirt plates an additional frictional pull is imposed. This pull is given by the formula on page 12.

Chain selection by section method **2.5**

An alternative method of calculating the chain pull is by the section method, i.e. a complete calculation for each section that has a different load condition. This is necessary on conveyors and elevators where the load is not constant over the whole of the conveyor, where the chains slide on their plate edges or where changes in direction occur.

It is well to remember that for uniformly loaded chains, there is a progressive increase in load pull from the zero position A to the maximum at D. This is illustrated graphically in Fig 2.3 where the vertical distances represent load pull occurring at particular points in the circuit, the summation of which gives the total load pull.

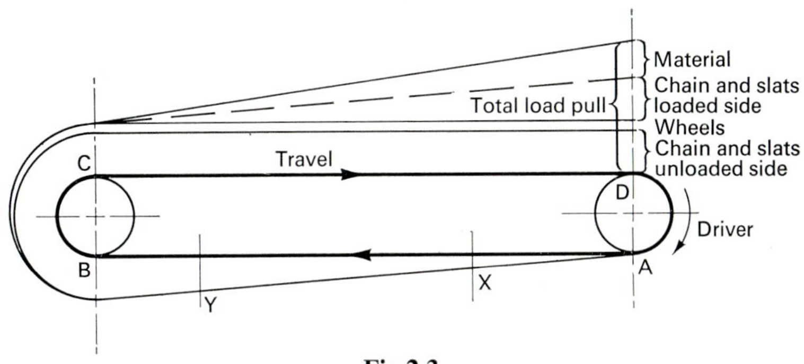

Fig 2.3

Thus if it is imagined that the chains are 'cut' at position X then there will be a less load pull or tension at this position than at Y. This fact is significant in the placing of caterpillar drives in complex circuits (Chapter 15) and also in assessing tension loadings for automatic take-up units (Chapter 16).

HORIZONTAL CONVEYOR CHAINS RUNNING ON ROLLERS

Referring to Fig 2.4 showing a diagrammatic side elevation of a simple slat conveyor, where the loaded and return strands are supported by running the chain rollers on guide tracks.

The maximum chain pull will occur at the driver position D, and minimum pull (theoretically zero) directly after the drive at position A. To assess the total chain load pull, the calculation is commenced at position A. The total chain pull comprises the following component load pulls.

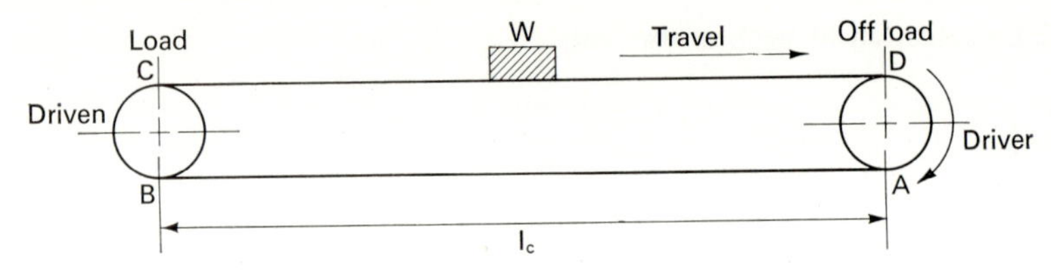

Fig 2.4

Section	*Load pull required*
A—B	To move the chains and slats in the unloaded strands.
B—C	To turn the tail wheels against the reaction load of the unloaded and loaded strands.
C—D	To move the chains and slats on the loaded strands. To move the load.

Then:—

Section	*Load pull required*	
A—B	$\mu_R l_C w$	(1)
B—C	It is common practice to apply a percentage figure of the load pull occurring at entry onto the wheels. This is normally taken as 5% so that the load pull becomes:—	
	$0{\cdot}05\,\mu_R l_C w$	(2)
C—D	$\mu_R l_C w$	(3)
	$\mu_R W$	(4)

Total load pull (lbf) = (1) + (2) + (3) + (4) which simplified

$$= \mu_R(2{\cdot}05\ l_C w + W) \qquad (5)$$

Where W = Weight of material on conveyor (lb)

w = Weight of chain(s) and conveying attachments (lb/ft)

μ_R = Overall coefficient of friction for chain rolling (see Assessment of chain roller friction Section 2.7)

l_C = Conveyor centres (ft)

HORIZONTAL CONVEYOR—LOADED STRAND ROLLING RETURN STRAND SLIDING

In this case the appropriate sliding coefficient of friction between the slats or chain and the supporting members should replace the chain roller coefficient in Formula (1) above. If this coefficient of sliding friction is designated μ_S then the total load pull (lbf)

$$= 1{\cdot}05\mu_S l_C w + \mu_R(l_C w + W)$$

HORIZONTAL CONVEYOR—LOADED STRAND ROLLING
RETURN STRAND UNSUPPORTED

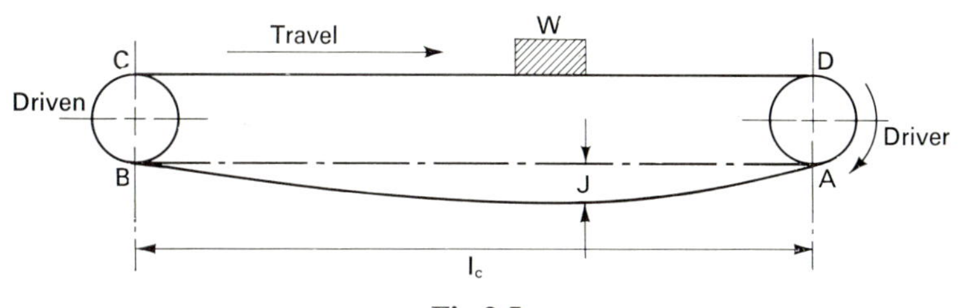

Fig 2.5

This can only be successfully applied to comparatively short centre systems as relatively high catenary load pulls are induced.

$$\text{Catenary pull in return strand AB (lbf)} = \frac{l_C{}^2 w}{8J} + wJ$$

$$\text{Total load pull (lbf)} = 1{\cdot}05\left(\frac{l_C{}^2 w}{8J} + wJ\right) + \mu_R l_C w + \mu_R W$$

Where J = Chain sag (ft)

HORIZONTAL CONVEYOR—LOADED STRAND ROLLING
RETURN STRAND ON IDLERS

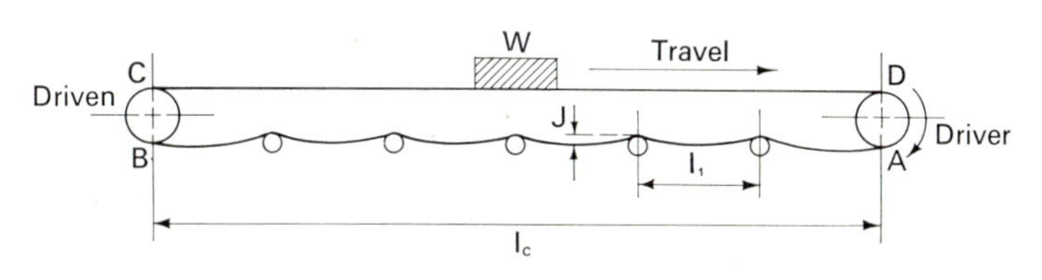

Fig 2.6

Here the chains form a catenary between a series of spaced idler rollers as shown in Fig 2.6. Calculations are similar to the conveyor with return strand unsupported, the formula for total load pull (lbf)

$$= 1{\cdot}05\left(\frac{l_I{}^2 w}{8J} + wJ\right) + \mu_R l_C w + \mu_R W$$

INCLINED AND VERTICAL CONVEYORS—CHAINS ROLLING

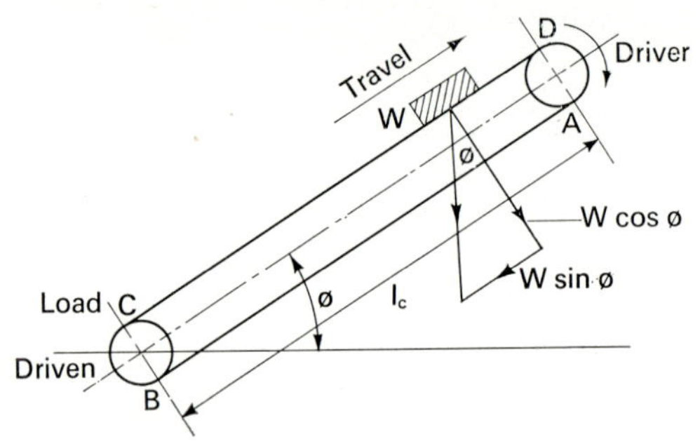

Fig 2.7

When a body of weight W (lb) rests on an incline, as Fig 2.7, then this weight can be resolved into two components:—

Parallel to the plane $= W \operatorname{Sin}\phi$
Perpendicular to the plane $= W \operatorname{Cos}\phi$

To move the load up the plane the force (F) required

$= W\,(\mu_{S1}\operatorname{Cos}\phi + \operatorname{Sin}\phi)$

Where μ_{S1} = Coefficient of friction between the load and the plane
ϕ = Conveyor incline (degrees)

This is precisely the condition when a chain is required to support and move a load up an incline; in this instance μ_R is the overall coefficient of friction of the chain rolling along the plane. Similarly the effort required to pull the return chain down the slope is given by the formula

$$P = W(\mu_R \operatorname{Cos}\phi - \operatorname{Sin}\phi).$$

Here it should be noted that if $\operatorname{Sin}\phi$ is greater than $\mu_R\operatorname{Cos}\phi$ then the effort required will be negative, i.e. the chain will roll freely down the slope. For inclines with tangents of less than μ_R a certain effort will be necessary to move the return strand of chain down the incline.

Referring to Fig 2.7 with chain running on its rollers.
Commencing at position A.

Section	*Load pull required*	
A—B	$(\mu_R\operatorname{Cos}\phi - \operatorname{Sin}\phi)l_Cw$	(1)
B—C	$0{\cdot}05(\mu_R\operatorname{Cos}\phi - \operatorname{Sin}\phi)l_Cw$	(2)
C—D	$(\mu_R\operatorname{Cos}\phi + \operatorname{Sin}\phi)l_Cw$	(3)
	$(\mu_R\operatorname{Cos}\phi + \operatorname{Sin}\phi)W$	(4)

Therefore chain pull (lbf)

$= 1{\cdot}05(\mu_R\operatorname{Cos}\phi - \operatorname{Sin}\phi)l_Cw + (\mu_R\operatorname{Cos}\phi + \operatorname{Sin}\phi)(W + l_Cw)$

Where:—

W = Weight of material on conveyor (lb)
w = Weight of chains and conveying attachments (lb/ft)
l_C = Conveyor centres (ft)
μ_R = Overall coefficient of friction for chain—chain rollers running on track
ϕ = Conveyor incline (degrees)

If the effort in equations 1 and 2 is negative then this is ignored when calculating the chain pull and the total chain pull is given by the summation of equations 3 and 4.

Therefore chain pull $= (\mu_R \text{Cos}\phi + \text{Sin}\phi)\ (W + l_C w)$

Later the negative quantities are used when calculating horsepowers.

TRACKED BENDS

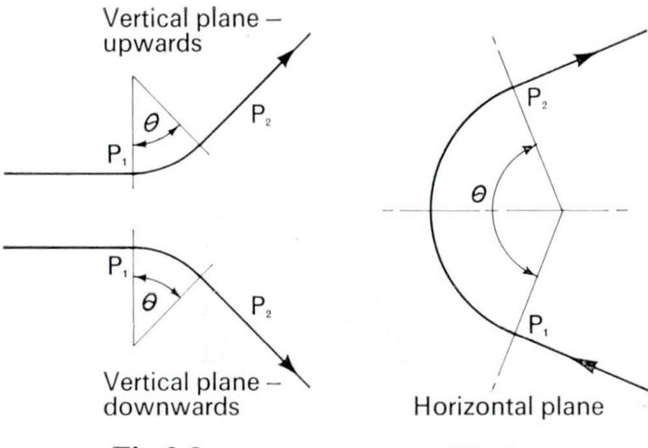

Fig 2.8 **Fig 2.9**

Where chains are guided around curves there is an inward reaction pressure acting in the direction of the curve centre. This applies whether the curved tracks are in the vertical or horizontal planes, and, relative to the former, whether upwards or downwards in direction. The load pull effect resulting from the chain transversing a curved section, even if this be in the vertical downward direction, is always considered as a positive value, i.e. serving to increase the chain load pull.

An analogy is a belt on a pulley whereby the holding or retaining effect depends upon the extent of wrap-around of the belt, and friction between the belt and pulley.

Similarly there is a definite relationship between the tension or pull in the chain at entry and exit of the curve. Referring to Figs 2.8

and 2.9 this relationship is given by:—

$$P_2 = P_1 e^{\mu\theta}$$

Where P_1 = Chain pull at entry into bend (lbf)
P_2 = Chain pull at exit from bend (lbf)
e = Naperian logarithm base (2·718)
μ = Coefficient of friction between chain and track
θ = Bend angle (radians)

The above formula applies whether the chain is tracked via the chain rollers or by the chain plate edges bearing on suitable guide tracks.

Table 2.8 gives values of e^x, where x is the product of $\mu\theta$. It is presented in this manner because of the many possible combinations of values of μ and θ.

Since high reaction loadings can be involved when negotiating bend sections it is usually advisable to check the resulting roller loading. This can be done from the following formula where L is the load per roller due to the reaction loading at the bend section.

$$L(\text{lbf}) = \frac{P_2(\text{lbf}) \times \text{Chain pitch (in)}}{\text{Chain curve radius (in)}}$$

There is a minimum radius which the chain can negotiate without fouling link plate edges (see Fig 2.10). Relevant minimum radii against each chain series can be obtained from chain catalogue.

c = Minimum clearance (in) must be 0·05in for small chains and up to 0·20in for chains over 36,000 lbf breaking load.
r = Minimum radius (in) which will vary according to pitch, roller diameter and plate depth.

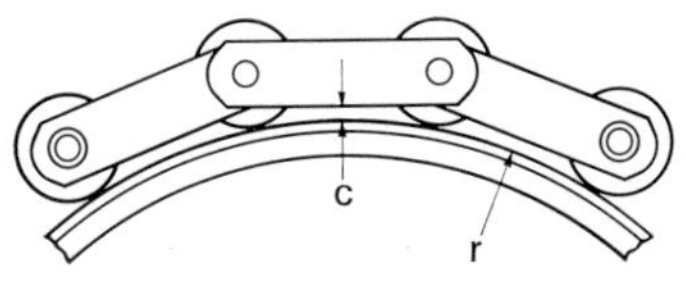

Fig 2.10

2.6 Horsepower assessment

FORMULA METHOD

On straightforward types of conveyors and elevators the approximate horsepower required to drive can be obtained by using the following formulae:—

Fig 2.11 Horsepower formulae

Material sliding

Horizontal, to 8°

Horsepower

$$hp = \frac{S \times (Wf_4 + W_A f_3)}{33{,}000}$$

Over 8°, to vertical

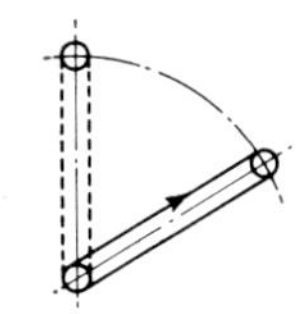

Horsepower

$$hp = \frac{S \times (Wf_4 + W_A f_5)}{33{,}000}$$

Material carried

Horizontal, to 8°

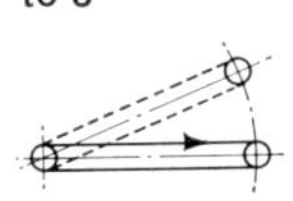

Horsepower

$$hp = \frac{S \times (Wf_2 + \frac{W_A}{2} f_3)}{33{,}000}$$

Over 8°, to vertical

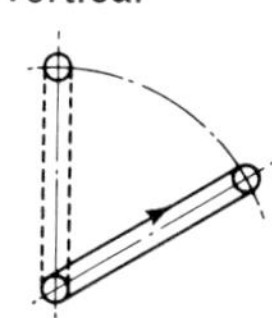

Horsepower

$$hp = \frac{S \times (Wf_2 + \frac{W_A}{2} f_5)}{33{,}000}$$

Where:—

W = Weight of material on conveyor (lb)
W_A = Actual total weight of chains and conveying attachments (lb/ft)
S = Chain speed (ft/min)
f_2 = See Table 2.6
f_3 = Sec Table 2.7
f_4 = See Table 2.2
f_5 = See Table 2.9

SECTION METHOD

This method can be used on all types of conveyor or elevator. The general formula for horsepower being:—

$$hp = \frac{\text{Unbalanced chain pull (lbf)} \times \text{Chain speed (ft/min)}}{33{,}000}$$

For all horizontal conveyors the unbalanced chain pull will be identical to the total load pull calculated when determining the strength of chain required. For elevators and conveyors with inclines then the effect of chain and load moving down the incline must be taken into consideration as this reduces the horsepower required to drive.

Using the examples from Section 2.5 then the horsepower for these is as follows.

Horizontal conveyor—Chains running on rollers

$$hp = \frac{\mu_R(2{\cdot}05\ l_C w + W) \times \text{Chain speed (ft/min)}}{33{,}000}$$

Horizontal conveyor—Loaded strand rolling
Return strand sliding

$$hp = \frac{[1{\cdot}05\mu_S l_C w + \mu_R(wl_C + W)] \times \text{Chain speed (ft/min)}}{33{,}000}$$

Horizontal conveyor—Loaded strand rolling
Return strand unsupported

$$hp = \frac{[1{\cdot}05\left(\frac{l_C{}^2 w}{8J} + wJ\right) + \mu_R(wl_C + W)] \times \text{Chain speed (ft/min)}}{33{,}000}$$

Horizontal conveyor—Loaded strand rolling
Return strand on idlers

$$hp = \frac{[1{\cdot}05\left(\frac{l_I{}^2 w}{8J} + wJ\right) + \mu_R(wl_C + W)] \times \text{Chain speed (ft/min)}}{33{,}000}$$

Inclined and vertical conveyors—Inclines with tangents less than μ_R (chain roller friction)

$$hp = \frac{[1{\cdot}05(\mu_R \text{Cos}\phi - \text{Sin}\phi) l_C w + (\mu_R \text{Cos}\phi + \text{Sin}\phi)(wl_C + W)] \times \text{Chain speed (ft/min)}}{33{,}000}$$

Inclined and vertical conveyors—Inclined with tangents greater than μ_R

$$hp = \frac{[1{\cdot}05(\mu_R \text{Cos}\phi + \text{Sin}\phi)(wl_C + W) - (\mu_R \text{Cos}\phi - \text{Sin}\phi) l_C w] \times \text{Chain speed (ft/min)}}{33{,}000}$$

2.7 Assessment of chain roller friction

In conveyor calculations the value of the coefficient of friction of the chain roller moving on the tracks has a considerable effect on chain selection. Table 2.6 gives values relating to the chains shown in Table 1.1. For other chains it is necessary to be able to assess their overall friction coefficient reasonably accurately, this is done as follows.

When a body of weight W (lb) is caused to move along a horizontal plane by the application of a force F parallel to the plane as shown in Fig 2.12 then by established laws

$$F = W\mu_{SI}$$

Where μ_{SI} = Coefficient of sliding friction between body and plane.

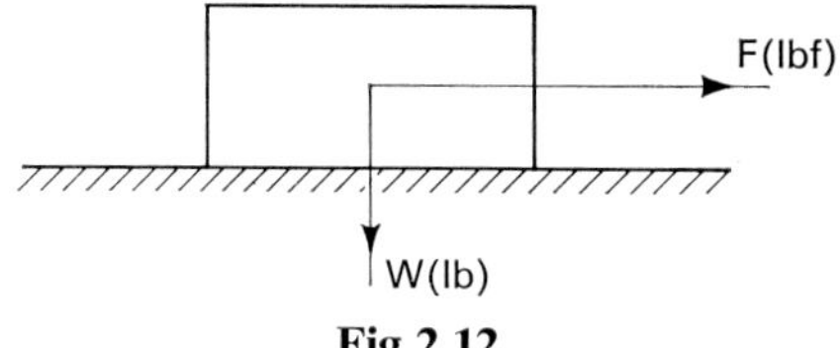

Fig 2.12

A similar condition of sliding friction exists between the roller bore and bush periphery as shown in Fig 2.13. For conditions of good lubrication a value μ_{S4}, 0·15 is reasonable and for poor lubrication approaching the unlubricated state, a value of 0·25 should be used.

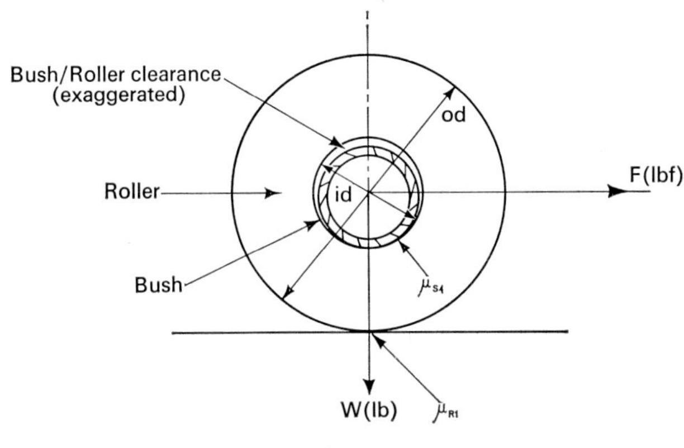

Fig 2.13

$$\text{Resistance to sliding} = \frac{\mu_{S4} \times \text{Roller inside radius (in)}}{\text{Roller outside radius (in)}}$$

$$= \frac{\mu_{S4} \times \text{id}}{\text{od}}$$

In addition to the sliding friction, resistance to motion is also caused by rolling friction. Experiments show that the values of the rolling friction (μ_{R1}) on surfaces which are related to conveyor work can vary between 0·002 and 0·005 for a steel roller running on a rolled steel track. Since operating conditions are normally not ideal it is usual to take the higher value 0·005. This figure applies to conditions at the periphery and must be related to the diameter of roller under consideration.

$$\text{Resistance to rolling} = \frac{\mu_{R1}}{\text{Roller radius (ft)}}$$

$$= \frac{2 \times 12 \times \mu_{R1}}{\text{Roller outside diameter (in)}}$$

$$= \frac{24\mu_{R1}}{\text{od}}$$

Overall coefficient of friction of chain μ_R = Resistance to rolling + resistance to sliding.

$$\text{Therefore } \mu_R = \frac{24\mu_{R1}}{\text{od}} + \frac{\mu_{S4} \times \text{id}}{\text{od}}$$

$$= \frac{24\ \mu_{R1} + (\text{id} \times \mu_{S4})}{\text{od}}$$

If average values of $\mu_{R1} = 0{\cdot}005$ and $\mu_{S4} = 0{\cdot}15$ are assumed this can be simplified to

$$\mu_R = \frac{0{\cdot}12 + 0{\cdot}15\text{id}}{\text{od}}$$

The above formula is applicable to any roller, but in the case of a wheel or outboard roller having ball, roller or needle bearings (Fig 2.14) the dimension id would be taken as the mean diameter of the bearing and μ_{S4} from 0·0025 to 0·005, the latter being assumed to apply in most conditions.

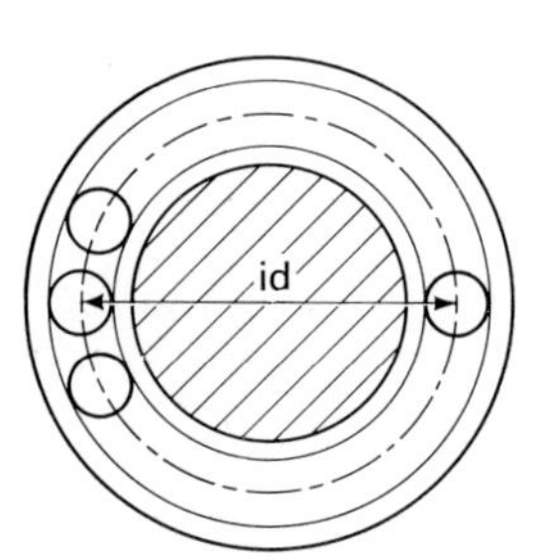

Fig 2.14

3

Slat Conveyors

Description and chain type 3.1

The application of slat conveyors is extensive throughout industry for conveying materials and equipment. They can also be used as travelling work tables.

Slats are normally attached to two chains by K attachments, the slats being either almost continuous with minimum gap between or spaced at larger intervals (see Chapter 5 for overlapping slat types); slat material is selected to suit the load being handled. It is usual for the chain to both move and carry the load; however outboard rollers can be employed if higher loads warrant this addition. Various methods of slat and chain arrangements are used depending on the weight and type of material being handled. Examples of these are shown diagrammatically as follows:—

Fig 3.1 This is a common type of light/medium duty slat conveyor using wood slats bolted to K attachments.

Fig 3.2 Here pressed steel slats bolted to K attachments give a light weight but rigid slat assembly. A cover flat on the loaded and return strands guards the chain.

Fig 3.3 In this assembly skirt plates prevent material fouling the chain. Skirt plates are also used on high level conveyors to ensure that material cannot fall from the conveyor.

Fig 3.4 and Fig 3.5 In these arrangements outboard rollers support the load leaving the chains to contend only with the chain pull.

Fig. 3.6 This illustrates a slat conveyor where the bars supporting the load are mounted through hollow bearing pins.

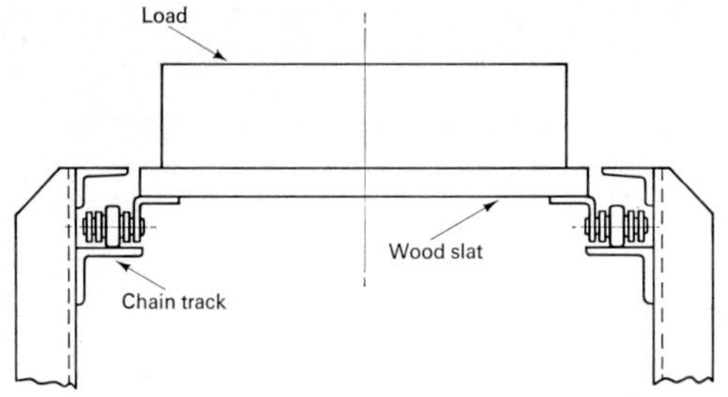

Fig 3.1

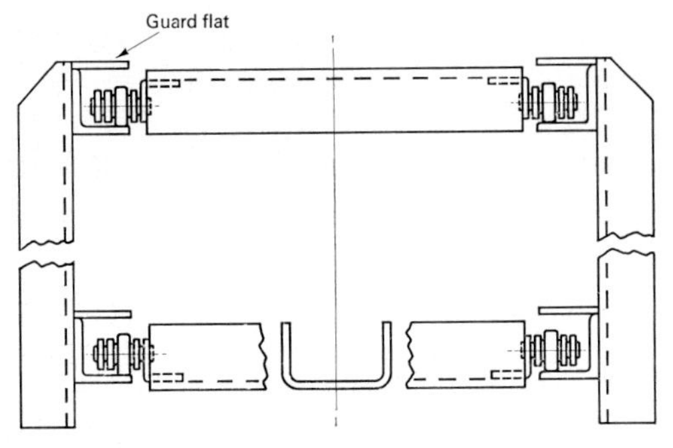

Fig 3.2

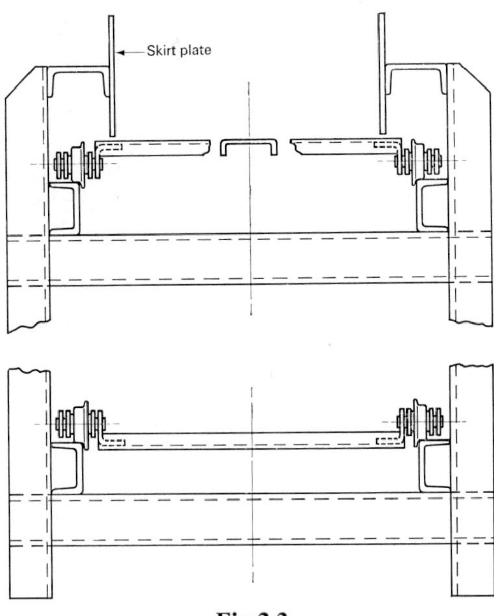

Fig 3.3

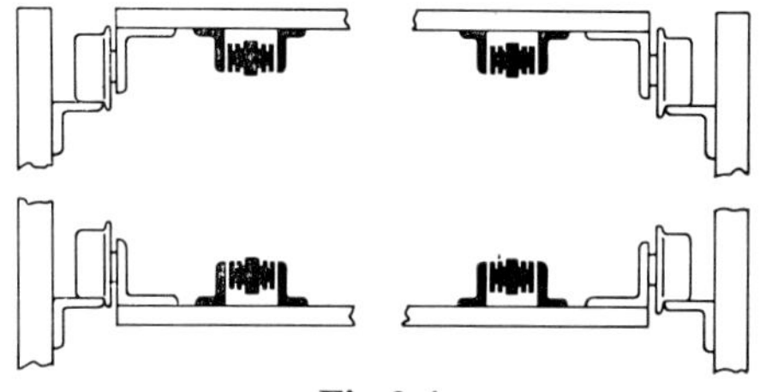

Fig 3.4

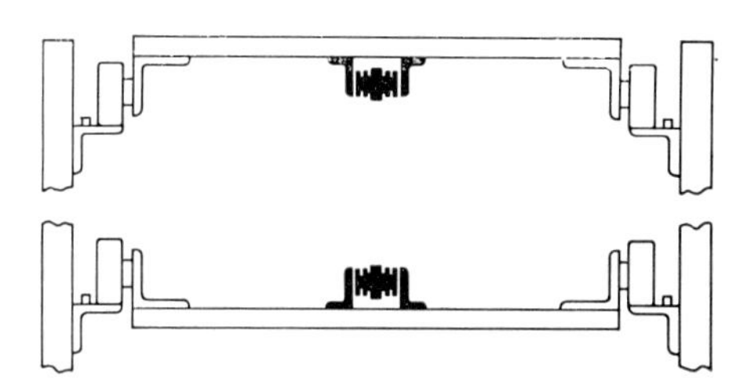

Fig 3.5

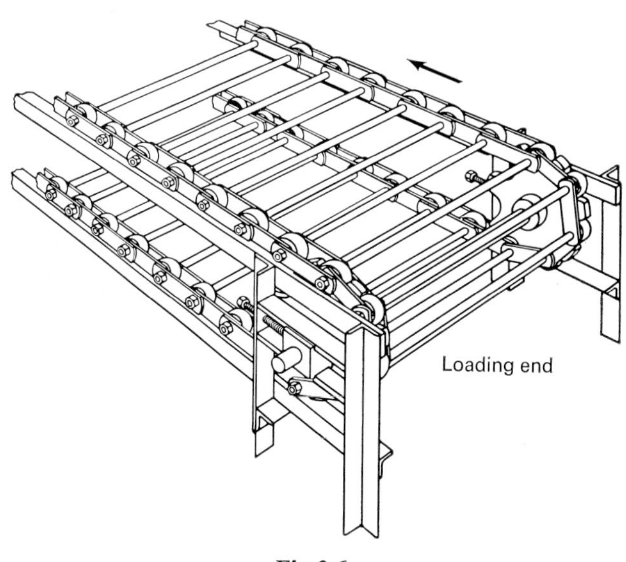

Fig 3.6

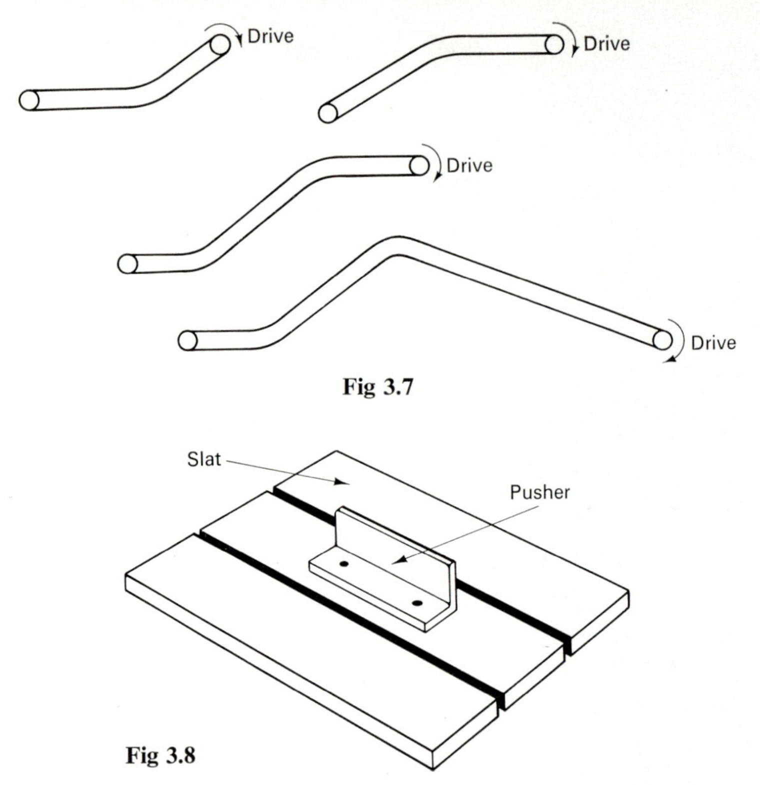

Fig 3.7

Fig 3.8

Horizontal conveyors and inclined conveyors with or without horizontal sections form the majority of slat conveyors. The introduction of inclines as shown in Fig 3.7 considerably broadens the application of slat conveyors. Care is necessary when deciding upon the conveyor incline, if the angle is too steep then the unit being conveyed will slip back down the conveyor. To overcome this the slats can be fitted with pushers spaced at intervals along the conveyor as shown in Fig 3.8. The slats should be jig-drilled to ensure uniformity and interchangeability.

Either hollow or solid pin chains may be used and generally these have K attachments. Mild steel or cast iron rollers are usually adequate but where heavy roller loading is involved then roller material should be considered in relation to permissible bush/roller bearing pressure (see Chapter 2). Cast iron wheels with 8 to 12 cast form teeth are suitable for normal applications.

If the load being conveyed requires drying then this can be achieved by using mesh slats or alternatively a wire belt. Endless wire mesh belts should be fitted between two strands of chain so that the onus of driving is removed from the belt. This is best achieved by using staybars as shown in Plates 1 and 2; this ensures that the belt centre line is coincident with that of the chain pitch line thus avoiding any tendency for the belt to stretch and malform when moving round the wheel. The limiting factors affecting staybar strength are as follows.

The extreme fibre stress should not exceed 10,000 lbf/in² for mild steel staybars.

The maximum slope does not exceed 0° 30′ (0·0087 Radians).

EXAMPLE

Calculate the maximum central load for a shouldered steel staybar 0·5 in diameter fitted between two chains at 15 in centres.
In calculations a staybar is normally taken as a simply supported beam.

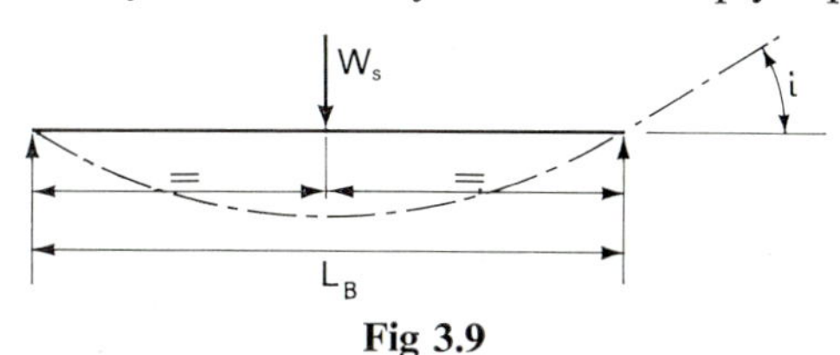

Fig 3.9

$$\text{Maximum slope} = i = \frac{W_S L}{2EI} \text{ (Radians)}$$

Where W_S = Weight of material on staybar (lbf)
L_B = Length of beam (in)
E = Modulus of elasticity (30×10^6 for Steel)
I = Moment of inertia (in⁴)

$$\text{Therefore } W_S = \frac{0{\cdot}0087 \times 2 \times 30 \times 10^6 \times 0{\cdot}003}{15} = 105 \text{ lbf}$$

$$\text{Also maximum extreme fibre stress } = f_S = \frac{My}{I}$$

Where M = Bending moment (lbf in)
y = Distance from neutral axis to extreme fibre (0·25 in)

$$\text{Therefore } M = \frac{10{,}000 \times 0{\cdot}003}{0{\cdot}25} = 120 \text{ lbf in}$$

$$\text{Now for a simply supported beam maximum bending moment} = \frac{W_S L_B}{4}$$

$$\text{Therefore } W_S = \frac{120 \times 4}{15} = 32 \text{ lbf}$$

Then the maximum central load is 32 lbf.

Alternatively, K attachments and support bars as shown in Fig 3.10 can be used.

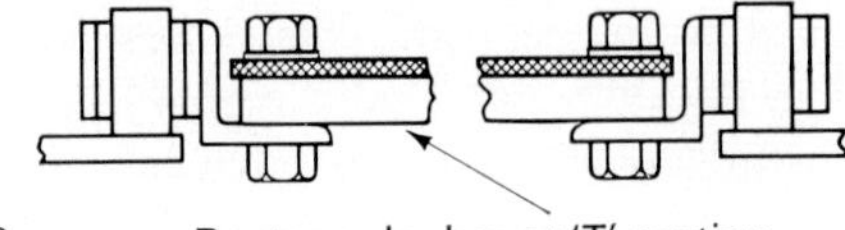

Fig 3.10 Rectangular bar or 'T' section

With this arrangement it is necessary to fix the belt to the support bars, the pitch of the bolts being about 8 in to 14 in. When the belt is not supported at every chain pitch it is necessary to incorporate pulleys on head and tail shafts to support the belt as it moves round the wheels.

On small wire belt conveyors use may be made of the hollow bearing pin feature available with transmission chain.

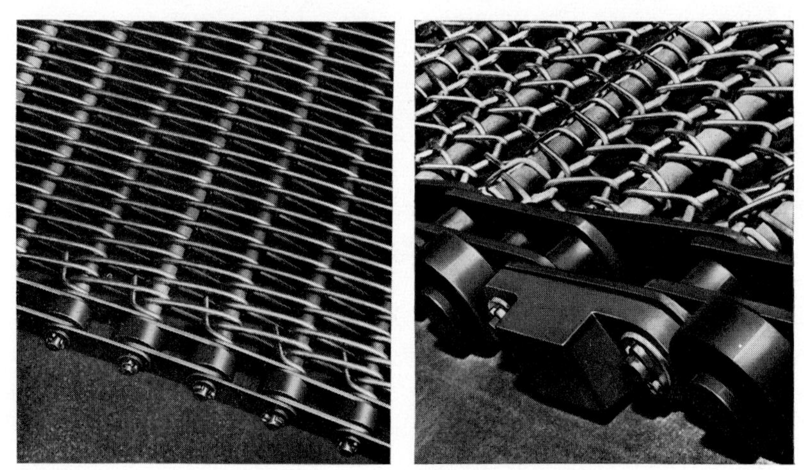

Plate 1 **Plate 2**

3.2 Chain pull calculation and horsepower assessment

This can be carried out by the formula or section method described in Chapter 2. The formula method is used in the following example.

EXAMPLE

It is required to select chains for a two strand slat conveyor carrying cartons 18 in cube each weighing 100 lb spaced at 30 in intervals on a 120 ft horizontal conveyor. The slats, assembled every pitch are 27 in long by 5·75 in wide and weigh, with fixing bolts, 3 lb each. The chains are 6 in pitch with K attachments one side every pitch; run on guide tracks both on the loaded and unloaded strands. Conditions are clean with occasional lubrication. Speed is 30 ft/min.

Total weight of material on conveyor $W = \frac{120}{2{\cdot}5} \times 100 = 4{,}800$ lb

Total weight of attachments (on loaded and unloaded side)
$=120\times3\times2\times2=1{,}440$ lb

Estimated total weight of chains and conveying attachments

$$W_E=2\times1{,}440=2{,}880 \text{ lb}$$

Following the procedure as described under Section 2.4 the preliminary chain selection can be made from the formula in Fig 2.1

$$\begin{aligned}\text{Chain pull} &= (W\,f_1)+(W_E\,f_1)\\ &= (4{,}800\times0{\cdot}22)+(2{,}880\times0{\cdot}22)\\ &= 1{,}690 \text{ lbf}\end{aligned}$$

From Table 2.3 a safety factor of 9 appears reasonable.

$$\begin{aligned}\text{Chain strength required} &= \frac{1{,}690\times9\times1}{2}\\ &= 7{,}605 \text{ lbf}\end{aligned}$$

Because of the comparatively long centre distance and therefore possibility of structural 'snaking' solid pin chains are preferred thus two strands of 7,500 lbf chain are provisionally selected. The exact chain weight can now be established.

Weight of chain with plain roller $=1{\cdot}62$ lb/ft.
Weight of bent over K2 attachments $=0{\cdot}31$ lb each

$$\begin{aligned}\text{Total weight of chain} &= 120\times2\times2(1{\cdot}62+2\times0{\cdot}31)\\ &= 1{,}075 \text{ lb}\end{aligned}$$

$$\begin{aligned}\text{Total weight of chains and conveying attachments} &= 1{,}075+1{,}440\\ &= 2{,}515 \text{ lb}\end{aligned}$$

Before proceeding with chain pull calculations it is now advisable to check the roller loading.

$$\text{Load per roller due to carton}=\frac{100}{6}=16{\cdot}66 \text{ lbf}$$

$$\text{Load per roller due to slat}=\frac{3}{2}=1{\cdot}5 \text{ lbf}$$

$$\text{Load per roller due to chain and K attachments}=\frac{1{\cdot}62+0{\cdot}31}{2}=0{\cdot}96 \text{ lbf}$$

$$\text{Total roller load}=19{\cdot}12 \text{ lbf}$$

As indicated by Table 2.5 a mild steel roller which is standard for the chain will be suitable, therefore final chain pull calculations can proceed.

By using the formula from Section 2.4

$$\text{Chain pull}=Wf_2+\frac{W_A}{2}(f_2+f_3)$$

$$=(4{,}800\times 0{\cdot}21)+\frac{2{,}515}{2}(0{\cdot}21+0{\cdot}21)$$

$$=1{,}536\ \text{lbf}$$

$$\text{Resultant safety factor}=\frac{2\times 7{,}500}{1{,}536}=9{\cdot}8$$

The running horsepower required to drive the conveyor is determined by the formula given in Section 2.6

$$\text{Horsepower}=\frac{30\times[(4{,}800\times 0{\cdot}21)+(2{,}515\times 0{\cdot}21)]}{33{,}000}$$

$$=1{\cdot}4$$

Plate 3 Cloth bale conveyor

The feed end of a conveyor handling bales of cloth from a dyeing and finishing plant to warehouse. At both head and tail ends, the conveyor is inclined for a short distance, necessitating formed cradle blocks on each slat. The conveyor is loaded manually, and off loaded by gravity. The number of slats has been restricted to the workable minimum in order to keep the overall weight, and consequent chain and power requirements, as low as possible. The slat ends project over the chains, but packings between slats and K attachments ensure sufficient clearance between top of roller and underside of slat for supports to be introduced to carry the chains on their rollers on the return strands.

Plate 4 Conveying heated aluminium sheets

Chain conveyors are ideal for handling hot material, standard chains being capable of withstanding temperatures up to 300°C. Use of special steels in manufacture can enable higher temperatures to be tolerated. In this application, two chains carry channel section slats with end cheeks to contain the aluminium sheets. Heat is dissipated by the large slat area and thus the temperature rise in the chains is minimised.

Plate 5 Wooden slat conveyor in a bakery

A simple wood slat conveyor is well illustrated, the slats being bolted between a pair of chains by bolts through K1 attachments at every pitch on one side of each chain. The wood slats are easy to clean and do not contaminate the loaves.

Plate 6 General goods conveyor

Installed at floor level in a railway warehouse to handle general merchandise from rail wagons to delivery vehicles, this conveyor employs wood slats bolted to K2 attachments riveted to all chain plates on one side of the chain. Part of the floor and three slats have been removed to clarify the details of the application.

Plate 7 Motor car assembly conveyor

Twin wood slat conveyors, set at the wheel track width of the vehicles conveyed and with chocks fore and aft of the vehicle wheels to prevent rolling, provide a controlled flow moving workbench along a vehicle assembly line. At certain stages, the conveyor climbs above floor level to permit easy access for operations to the car underbody. Absolute reliability is demanded of such conveyors, since breakdowns mean production loss. To this end, all guides and tracking are carefully built and aligned, slats are jig drilled for uniformity and speedy replacement, and each of the two sets of chains is matched for length. Certain operations can be performed on partially or fully completed vehicles by the use of only one slat conveyor when movement along a straight and horizontal path is required. Under this system one pair of wheels—say the off-side wheels—rests on the conveyor and the other wheels simply roll freely along the floor.

Plate 8 Heavy duty slat conveyor

The conveyor, carrying heavy unit loads from factory to warehouse, embodies metal slats to withstand wear and abrasion.
To permit the use of minimum thickness of slat material in the interests of economy in cost, weight and power needed to drive the conveyor and yet to obtain sufficient slat strength to sustain the heavy loads carried, the slats form a channel section.

Plate 9 Television set testing conveyor

Although basically a simple wood slat conveyor, this installation is of interest in that electrical and aerial sockets are built in at intervals so that the sets may be connected for testing as they are conveyed along the inspection bay. A variable speed drive unit matches the conveyor speed to the rate of production flow. Slats are fixed to K attachments and can be withdrawn easily whilst the conveyor is in motion, permitting access for rectification of electrical faults with minimum delay.

Plate 10 Cloth roll conveyor

This conveyor embodies special slats curved to the radius of the cloth rolls. They centralise the load and promote smooth running.

Plate 11 Slat conveyor for hot coiled strip

Hot strip material at near red heat is successfully handled by the conveyor shown, using standard chain. The tubular slat assemblies embody strength and rigidity, and at the same time aid heat dissipation and keep the temperature rise in the chains to a minimum. Added rigidity is provided by staybars between the chains, bolted through outer link centres.

Plate 12 Aluminium billet transfer conveyor

This conveyor transports the billets to a pre-soaking oven. Although the chains do not enter the oven. Two chains are employed but, in this case, short slats are fitted to individual chains. Load carrying capacity is the reason for this arrangement.
In this conveyor each slat is supported on four easily renewable rollers; the chain rollers are required only to gear with the driving wheels.

Plate 13 Slat conveyor/elevator for foam rubber

This ingenious conveyor installation makes use of the natural elasticity and lightness of foam rubber to elevate the pieces by gripping them between two wood-slat conveyors.

Plate 14 Special pocketed slats

Bottles are carried through a washer in metal slats with pockets, the slats being fixed to two chains by K attachments. Chain materials on these applications vary according to the practice of individual machine manufacturers, i.e. stainless materials throughout or a combination of stainless bearing pins, bushes and rollers, with link plates and attachments in normal materials.

Plate 15 Bar apron conveyor

Chains are here used for bar apron conveyors forming assembly lines in a factory producing typewriters.

Plate 16 Cask drying conveyor

Two distinct conveyors are used in this ingenious system for the drying of casks and barrels, after steam cleaning and sterilising. The outer chains, with their multiple attachment arrangement to accommodate varying lengths and diameters, carry the casks. Inboard of these two chains is situated in the airbox, this is essentially a twin strand conveyor with retractable spouts. At the commencement of travel along the conveyor the spout is directed into the cask bung hole and remains there until off-loading. Closely mating slats provide an effective air seal.

Plate 17 Roller slat conveyor

Free rollers, mounted between chains on extended bearing pins or spigot pins, may take the place of normal slats. This type of conveyor is intended for handling rigid unit loads such as cartons, cases etc., enabling two distinct functions to be performed.

With chains running on support tracks and rollers free to rotate, the load can be held stationary with the conveyor still running or propelled manually or mechanically at a higher linear speed than the conveyor.

Approaching the feed-off point, the free rollers may run on to a raised bed, thus transferring support from the chains to the rollers themselves. The load will then travel at twice chain linear speed to aid transfer to a further conveyor system.

Plate 18 Escalator *(opposite)*

The escalator, or moving staircase, is in effect a special slat conveyor, each step being a triangulated slat. Each step is fitted between a pair of chains pivoting on extended bearing pins. On the opposite side of the chain, at coincidental points, are mounted outboard rollers, which run in a guide track. The trailing end of each step carries two rollers which run on separate tracks positioned within the chain. The position of these tracks ensure that the steps on which the passengers stand are always in the horizontal position as shown in **(c)**.

Accurate pitching of the steps is necessary to keep clearances to a minimum during negotiation of the curved track sections. This dictates the need for special length tolerances on the chains, the chains are also matched to run as a pair.

Plate **(a)** is a general view of the escalator, whilst **(b)** shows the tail end assembly, situated beneath the lower floor where a spring loaded automatic chain take-up is used. The large number of teeth in the wheels, essential for smooth running, is illustrated.

LINOS • KITCHEN CABINETS
TOY FAIR
AND SANTAS
Reindeer Sleigh
Merry Christmas

(a)

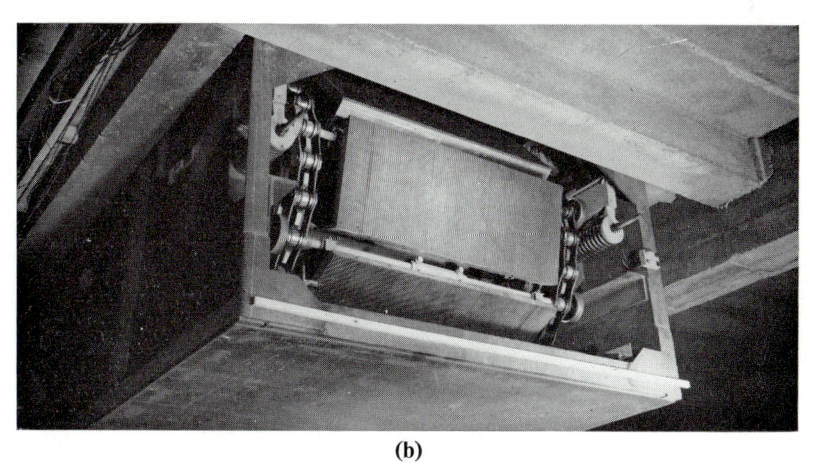

(b)

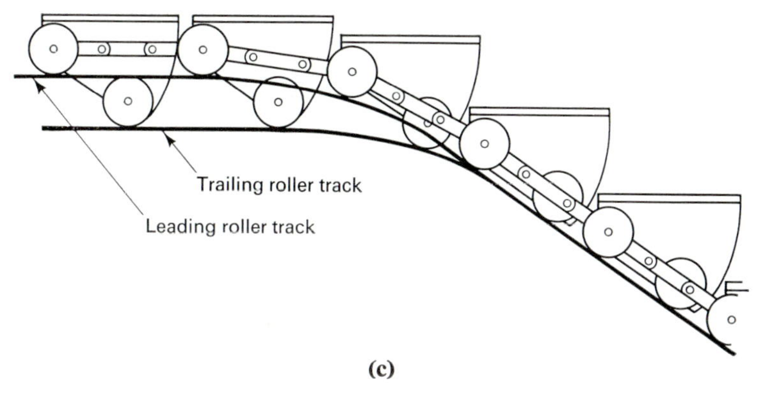
Trailing roller track
Leading roller track

(c)

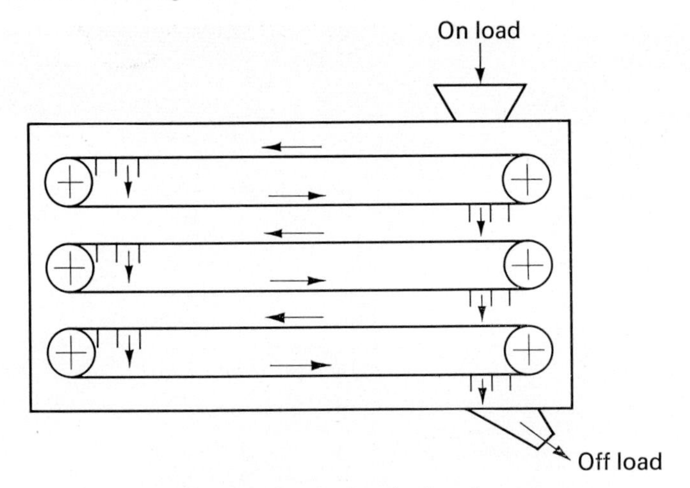

Drying chamber fitted with tipping slat conveyor

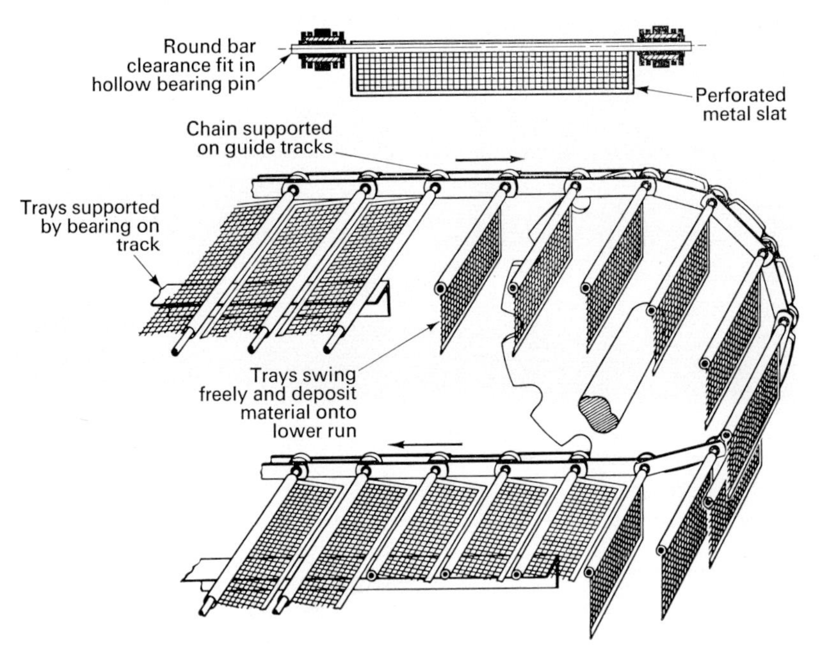

Plate 19 Tipping slat conveyor

The method is useful when loose, non-sticky or non-granular materials have to be disturbed to regularise drying or heating. One advantage is that both top and bottom strands of the conveyor are utilised. Operating speeds are normally slow, ranging from a few in/min to about 30 ft/min depending on the drying time required.

Plate 20 Mesh slat conveyor

A typical example is the vegetable and herb drying conveyor shown above. In this case the mesh slats are hinged and secured on staybars bolted through the hollow bearing pins of the chains.

4

Pusher Conveyors

4.1 Description and chain type

The pusher conveyor comprises basically a single chain operating as a propulsive medium only, with the load supported on an independent surface or track, and propelled by means of pusher attachments fitted to the chain as shown in Fig 4.1. Various examples of pusher conveyors are illustrated in Plates 21 to 25. Alternatively, a series of pusher conveyors can run in parallel to form a feed table for long bars, etc.

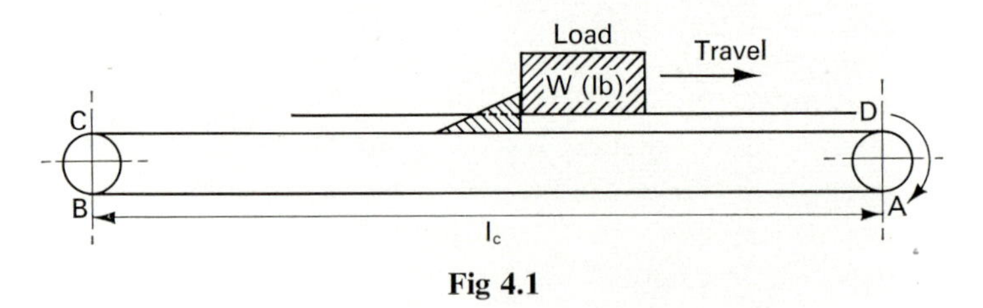

Fig 4.1

Solid pin chains should normally be used. On pusher conveyors the chain rollers are subject to reaction loading, and it may be desirable to introduce outboard rollers on both sides of the chain. Roller loadings due to reaction loads should always be checked against allowable values. An advantage of outboard rollers is that they are readily replaceable when worn.

There are various methods by which the pushers may be fixed to the chain, including K, F and S attachments, connecting links with extended pins or by drilling link plates.

If types as shown in Fig 4.2(a) and (b) are used, the chain rollers can provide support on both the loaded and return strands. If type (c) is employed without outboard rollers, the return strand cannot be supported, because of interference from the pusher attachment. The return strand can simply hang as a catenary between the head and tail wheels if space below is unrestricted and centres are less than about 20 ft. Usually however an outboard roller assembly of the type shown in Fig 4.3 is desirable. The spacing of outboard rollers however should not exceed about 5 ft; if pusher spacing required is in excess of this then additional outboard rollers should be introduced to give intermediate support.

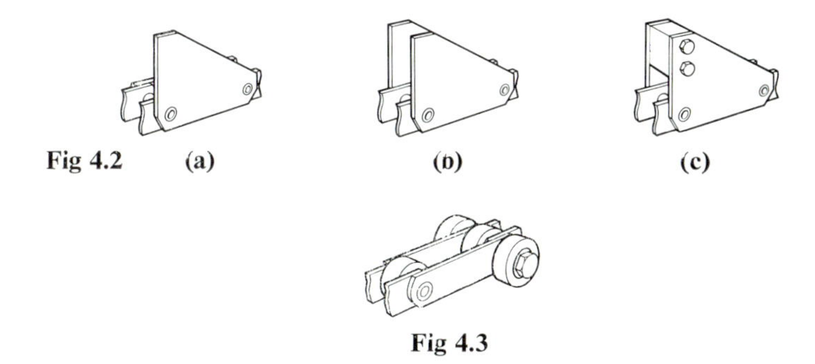

Fig 4.2 (a) (b) (c)

Fig 4.3

In some applications, it may be necessary for the chain to reverse motion past the load to be propelled, or for the load to overrun the chain as on mine-car pushers. In such cases, a tilting pusher attachment is used, the fulcrum point of the pusher being co-axial with the chain bearing pin. An example of this type is illustrated in Plate 25.

At positions immediately after the tailwheel and before the driving wheel, the load is respectively picked up and discharged. The chain track at these points is curved as shown in Fig 4.4 to ensure adequate chain support and smooth load pick-up.

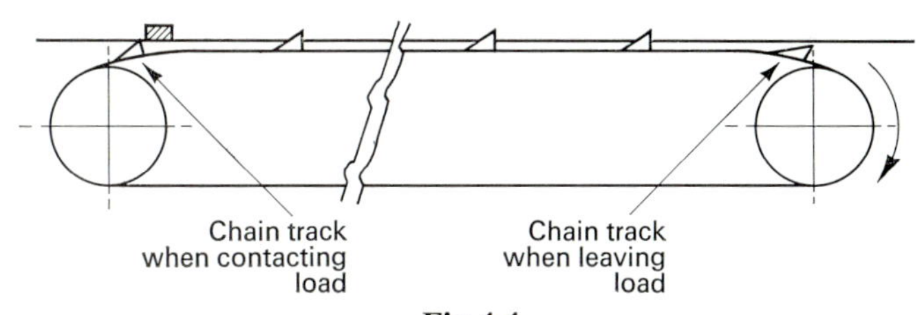

Fig 4.4

Cast iron wheels having 8 teeth are suitable for the majority of smooth running applications, i.e. no high 'pick-up' or accelerating loads. Where impulsiveness with high momentary loading is present steel wheels with an increased number of teeth, say 12 to 16, are recommended. Maintenance of the chains in correct adjustment is essential particularly if pick-up of the load occurs at appreciable speed.

In multi-strand systems each chain is driven via a common drive shaft whilst individual transverse centres may be considerable. To ensure equal load pull distribution, the shaft should be of such proportions and strength as to minimise torsional deflection. If a single prime mover is used this should preferably drive the headshaft at a position midway along its length.

4.2 Chain pull calculation

Smooth conditions

Referring to Fig 4.1

Total chain pull comprises the following component load pulls:—

Section	*Load pull required*
A—B	To move the chains and pushers on the unloaded side.
B—C	To turn the driven wheels.
C—D	To move the load.
	To overcome the reaction load at the pushers.
	To move the chains and pushers on the loaded side.

To review the loadings it can be assumed that the total load to be moved acts at one pusher position.

This is shown in Fig 4.5.

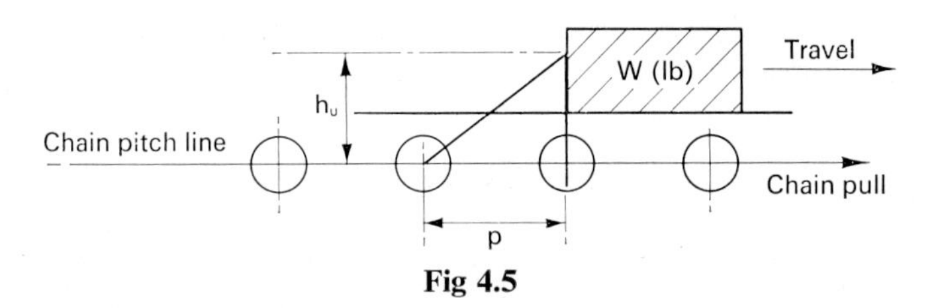

Fig 4.5

Where

W = Weight of material on conveyor (lb)

w = Weight of chain and conveying attachments (lb/ft)

l_C = Conveyor centres (ft)

μ_R = Overall coefficient of friction of chain

μ_{S2} = Coefficient of sliding friction between material and conveyor

p = Chain pitch (in)

h_U = Pusher height from chain pitch line (in)

Section	*Load pull required*		
A—B	To move the chain(s) and attachments	$=\mu_R l_C w$	(1)
B—C	To turn the driven wheels	$=0{\cdot}05\mu_R l_C w$	(2)

C—D To move the dead load $=\mu_S W$ (3)

To overcome the reaction load $=\mu_S W \frac{h_U}{p} \mu_R$ (4)

To move the chain(s) and pushers $=\mu_R l_C w$ (5)

Total chain pull = (1) + (2) + (3) + (4) + (5) which simplified

$$=\mu_S W\left(1+\frac{h_U \mu_R}{p}\right)+2{\cdot}05\mu_R l_C w$$

If the load is stationary and the chain pusher contact with the load occurs at an appreciable speed, account must also be taken of the impact force induced in the chain.

$$\text{Impact force} = mv \text{ (lbf)}$$

Where m = Mass of load to be moved

$$=\frac{W}{32{\cdot}2}$$

v = Chain speed at instant of impact (ft/sec)

Plate 21 Ice block pusher conveyor

Ice blocks slide along a smooth level surface, being propelled by pusher dogs fitted to a pair of standard chains. The chains are matched to run as a pair to keep the dogs in line across the conveyor; they run beneath the table surface, the dogs projecting through slots. Whilst conditions of operation are rather severe due to wetness, chains in standard materials are the best choice on an economic basis provided they are reasonably protected by a water repellent grease.

Plate 22 Log handling conveyor

Pusher dogs are mounted at intervals on each side of each chain with solid distance pieces between. Multiple chains are fitted and they elevate the logs, which are used in match manufacture, to the saw for cutting to length. The logs roll by gravity down the ramp in the foreground and on to the tracks, the chains being sited below the tracks to protect them from impact. Since each dog spans two pitches of chain, an elongated hole is provided in the trailing end of the dog plates to allow chordal pitch foreshortening to take place when negotiating the head and tail wheels.

Plate 23 Push bar elevator/conveyor for cartons

Another method of propulsion is to use staybars between two chains as shown. The point of contact between the push bar and article should not allow fall-back or create a tendency to trap against the supporting bed on the straight runs or where bends are to be negotiated. An alternative arrangement would be to use angle section pushers fitted to K or F attachments.

Plate 24 Multi-welding machine for motor car floor assemblies

The assemblies are placed in a wheeled jig which is conveyed round all four sides of the welder with pauses at various welding heads. The photograph shows the wheeled jig; the conveyor chain carrying the pusher dog is seen negotiating the chainwheel and about to contact the screwed buffer in the jig. Positional accuracy is most important on such an application as this, and is derived directly from the chain.

Plate 25 Semi-automatic handling plant for structural steelwork

Multiple chains are used in this plant for the transfer from a marshalling area to cross-feed powered rollers of structural steel sections for cutting off and drilling operations. The chains employ tilting pusher attachments which allow reversible movement, without fouling, and controlled pick-up of new sections.

5

Apron Conveyors and Feeders

5.1 Apron conveyors

DESCRIPTION AND CHAIN TYPE

The apron conveyor usually employs two endless chains to which overlapping slats, called aprons, are fixed. Various designs of aprons can be used, the object being to form a continuous travelling bed having the minimum amount of spillage. Some typical arrangements are shown in Fig 5.1 from which it will be seen that the leading edge of each apron is contoured over the trailing edge of the preceding one. If required, the aprons can be fitted with vertical end plates, this type being generally known as a pan conveyor or apron conveyor with closed pans.

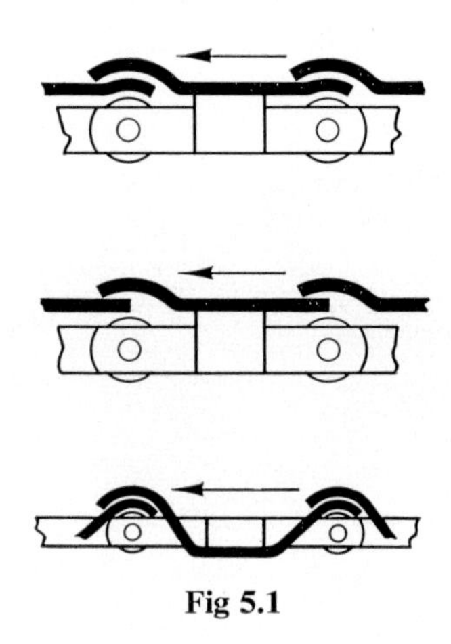

Fig 5.1

Apron and pan conveyors are extensively employed for handling bulk materials of both abrasive and mildly corrosive character, such as ores, sugar cane, gravel, clinker etc. Unit loads can also be handled. They operate successfully whether horizontally or inclined. Fig 5.2 shows various designs of aprons and pans for inclined working.

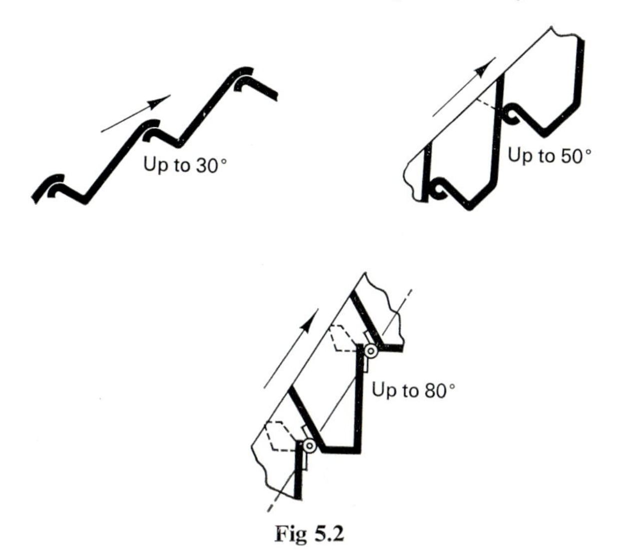

Fig 5.2

If stationary skirt plates (see Fig 5.3) are used on horizontal conveyors or those with only slight inclination then a deep bed of material may be carried successfully. Such skirt plates normally comprise part of the main conveyor structure. As a general rule, the height of the skirt plates should be 66% of the apron width, and the nominal height of the material (h) measured at the side of the conveyor should be about the same percentage of the skirt plate depth as shown in Fig 5.3. It must be appreciated that the pressure of material against the stationary skirt plates will increase the required load pull of the chains.

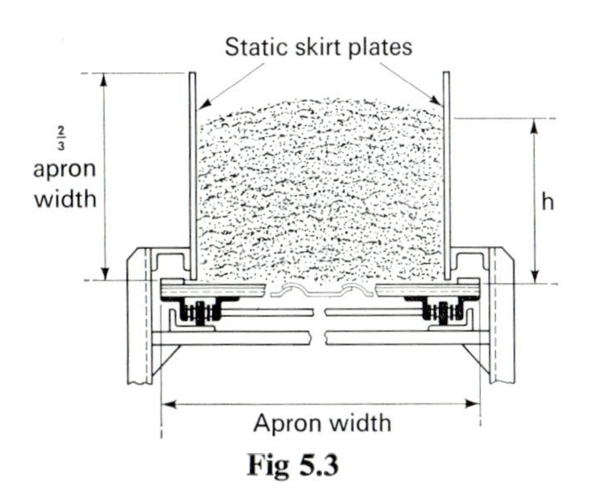

Fig 5.3

The feed-on of material to be conveyed, is often at right angles to the run of the conveyor. If this has a surge effect it can induce a side movement of the conveyor chains, which may require additional side guide control at this section. With pan conveyors, an effective method is to fit small diameter rollers at close spacing, these bearing against the end plate as shown in Fig 5.4. Alternatively, a plain reaction guide can be used.

To reduce apron deflection, with its adverse effect on the chain, staybars are sometimes employed as shown in Fig. 5.4 to provide a rigid chain and apron assembly. The longitudinal pitching of the staybars should coincide with that of the aprons.

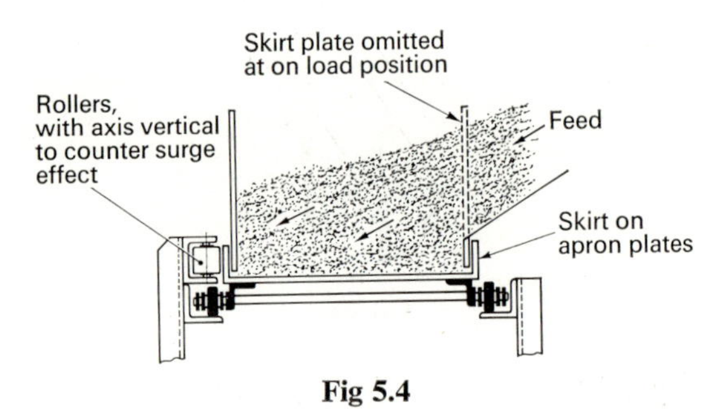

Fig 5.4

Another method of reducing slat deflection, particularly at the loading point, is to employ support rails as shown in Fig 5.5. These are set with a small clearance between the rail and the slat, the latter normally being fitted with wear shoes or rollers.

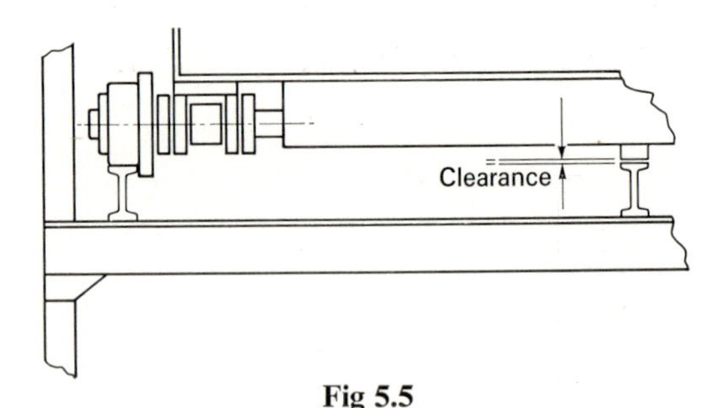

Fig 5.5

The chains are frequently mounted within the outer extremities of the apron plates, and cannot therefore be supported on the return run. In such cases, plain faced idler wheels support the apron and chains, these being spaced at about 4 ft to 6 ft intervals. Alternatively, the aprons may slide on a wood or metal track.

For all but the lightest duties solid bearing pin chains are used, in view of the arduous conditions of operation. The aprons or pans are secured to K attachments on one or both sides of the chain. Chain pitch is dictated by the width of the apron plates. The chain roller material should be of case-hardened mild steel, if abrasive loads are being handled; otherwise cast iron may be suitable. Cast iron wheels having 8 teeth are suitable for light duty. However, on arduous applications the wheels should be of steel, and the number of teeth increased to 12. Wheel tooth gapping is necessary if staybars are fitted through the chain plates.

CHAIN PULL CALCULATIONS

The formula is given in Chapter 2. If apron conveyors with static skirt plates are used then the formula will be as stated, but an additional load pull should be included to allow for the effect of side pressure of material against the skirt plates. This additional load pull is given by the expression:—

$$\frac{h_M^2}{F_M} \text{ (lbf/ft) of conveyor loaded length}$$

Where h_M = Height of material (in) (see Fig 5.3)
F_M = Factor depending on material handled.

Typical F_M values for material against steel plates are:—

Material	F_M *factor*
Cement clinker	8
Clay (compact)	48
Coal (bituminous dry fines and lumps)	30
Gravel, stone	8
Iron ore (crushed)	4
Sugar cane	70
Wood chips, pulpwood	48

Apron feeders 5.2

DESCRIPTION AND CHAIN TYPE

The apron feeder conveyor is similar to the apron conveyor already described. Material is normally received from a hopper or chute as shown in Fig 5.6. The conveyor usually operates horizontally at a slow speed over short distances, sometimes with an intermittent travel. The apron feeder conveyor provides a positive and regulated material feed to other conveyors or processes. They are frequently called to handle heavy bulk material in large pieces, often highly abrasive in character, such as rock, limestone, ores and so on.

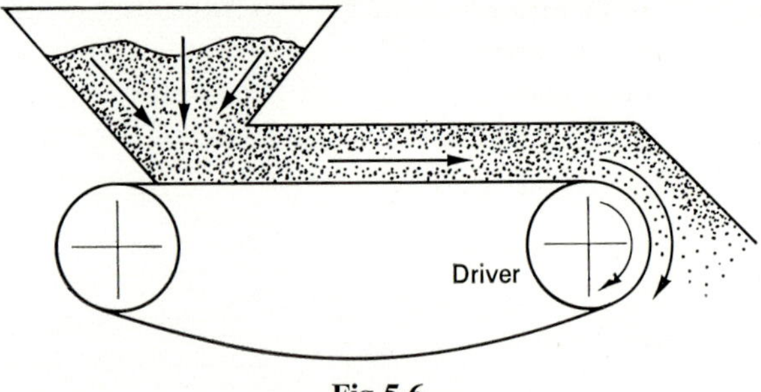

Fig 5.6

The method of direct feed may often entail the free dropping of material from an appreciable height on to the conveyor. In such circumstances all components, particularly the chains, have to be of extremely robust construction. If a layer of material is already on the feeder (when a further load is dropped) this can have a cushioning effect, and if possible the feeding cycle should be so arranged that this effect is deliberately timed. If the feed takes place down an angled section of the hopper, a series of suspended heavy cable link chains can be arranged in the hopper to act as a retarding device for the falling material, and so reduce the impact. A similar effect can be obtained by baffle plates.

The chains normally run on fixed guide tracks, but in order to absorb part of the impact at the on-loading points a section of track can be supported by springs or rubber cushion mountings. It may also be desirable to provide additional support tracks at the load point to limit deflection of the apron plates. On short centre distance feeders it is common practice to allow the return chain strands to hang in a free catenary as shown in Fig 5.6. This is permissible provided due account is taken of the catenary load pull; it is however preferable to provide some form of positive support such as idler support rollers.

The chains should invariably be of solid bearing pin type, with K attachments. Special attention must be paid to the method of fitting the attachment to the chains in order to afford adequate strength. Preferably integral, i.e. bent-over attachment chain plates should be used or alternatively fully welded K attachments.

The chain pitch is usually dictated by the width of the apron plates, along with the provision of wheels having an adequate number of teeth within the space available. The loading on individual chain rollers is also an important consideration; case-hardened mild steel rollers should be employed. Wheels having 12 teeth are to be preferred, since the material is usually abrasive and the duty arduous, steel wheels are required.

It is desirable to incorporate a torque limiting device or clutch at the headshaft arranged to operate when the torque applied subjects the chains to a load pull greater than 20% of their breaking strength.

CHAIN PULL CALCULATIONS

The formula is given in Chapter 2.

Bunker conveyors 5.3

DESCRIPTION AND CHAIN TYPE

The bunker conveyor for handling bulk materials is a further type of apron conveyor; the principle being shown in Fig 5.7. They are used for the storage and handling of many free flowing materials. Operation is at low speeds of 1 ft to 15 ft per minute, with manually operated control gate to regulate the flow.

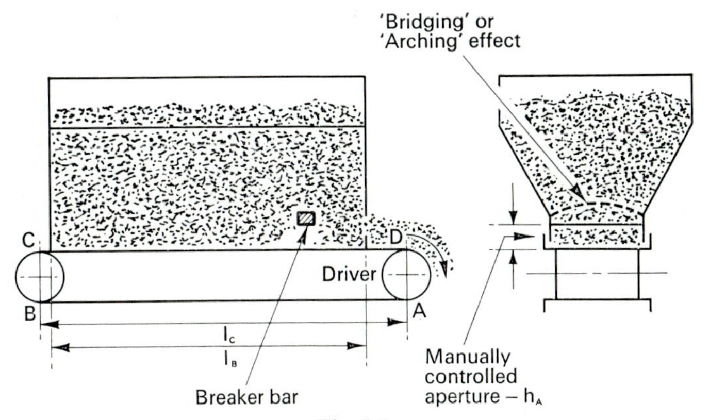

Fig 5.7

Above the top line of the gate, a robust breaker bar or similar device is often fitted at the point just prior to discharge. The object is to cause material segregation which aids discharge. The bunker may be 'open' so that material initially discharged into it, from lorries, trucks etc., impinge directly on to the conveyor or can incorporate baffle plates to reduce the effect of impact loading. Materials with difficult flow characteristics can also be handled but a degree of mechanical or manual agitation will be required to effect discharge.

Other features have already been covered under apron feeders.

CHAIN PULL CALCULATIONS

It is difficult to predict with accuracy the load pull requirements on this type of conveyor due to the bridging effect of the material coupled with the drag-out effect required. A method which gives reasonably satisfactory results in practice is described but must be regarded as only approximate.

Referring to Fig 5.7 the total load pull comprises the following component load pulls.

Section	*Load pull required*	
A—B	To move the chains and apron plates on the unloaded side.	(1)
B—C	To turn the driven wheels.	(2)
C—D	To move the chains and apron plates on the loaded side.	(3)
	To move an amount of material equal to aperture height (h_A) over the conveyor centres. Note an additional friction pull will be caused by the material sliding against the vertical side of the bunker.	(4)
	To 'drag-out' the material beneath the main body of material in the bunker.	(5)

Component load pulls (1), (2), (3) and (4) are calculated as shown in Chapter 2, it is the pull (5) on which further comment must be made, The proportion of weight of material in the bunker that is supported by the conveyor depends upon such variables as angle of bunker sides, repose angle of material and flow characteristics of the material. However, from a practical aspect it is assumed that the weight of material acting on the conveyor is represented by the area A, B, C, D as shown in Fig 5.8.

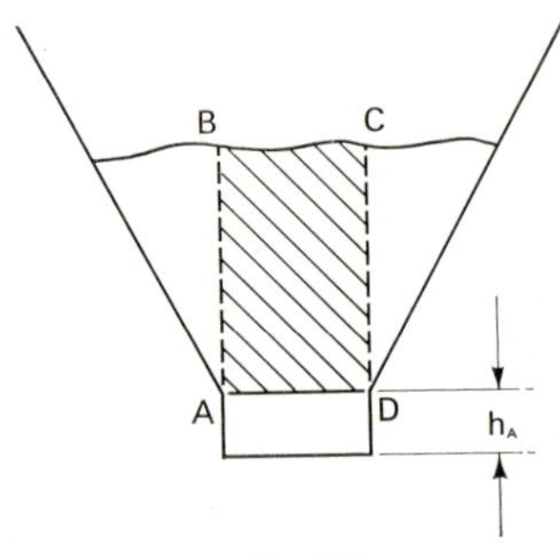

Fig 5.8

Therefore load pull (5) $= W_M \mu_{S3} l_B$

Where W_M = Weight of material supported by conveyor (lb)

μ_{S3} = Coefficient of friction of material sliding against itself

l_B = Length of bunker (ft)

Plate 26 Apron plate catenary conveyor

A modified form of apron conveyor employing comparatively large ball bearing outboard rollers at about 6 ft intervals. The high side plates give a good capacity, the installation shown handling 400 ton of cement clinker per hour at the low speed of 90 ft/min.

Plate 27 Underfloor apron pan conveyor

This conveyor handles aluminium sheet offcuts and the overlapping slats have integral raised sides. The slat overlap is at the leading edge so that on opening out when negotiating the wheels at the off-load end of the conveyor, the material cannot fall between the slats. The guide tracks immediately below the top strand of the conveyor are carried close to the wheels, giving well controlled lead-on of the chains.

Plate 28 Inclined swarf conveyor

Overlapping slats are employed running between fixed skirt plates, to minimise spillage. Change from incline to level is achieved by running the chains over curved guide tracks and, the chains being outboard, this applies also to the return run. High reaction loads at these curves make replaceable curved tracks a desirable feature.
The incline up which the swarf can be conveyed depends upon its friction on the slats, but the fitting of pusher dogs at intervals, as shown, increases the permissible slope quite considerably.

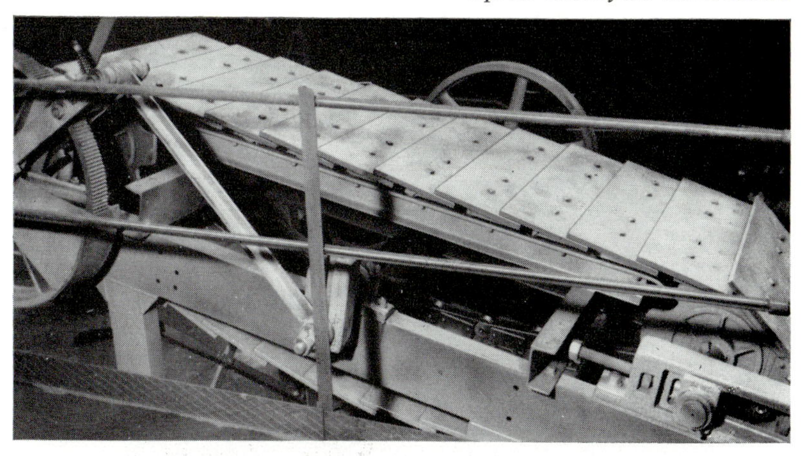

Plate 29 Apron feeder

Medium duty feeder with plates fitted to angled K attachments. The drive is via a countershaft and ratchet drive. The direction of travel relative to the slat overlap avoids trapping and spillage between the slats at the delivery end.

Plate 30 Limestone feeder

Two strands of 85,000 lbf breaking load conveyor chain carrying overlapping steel slats by means of K attachments riveted to the chain plates on this limestone apron feeder. The feed is intermittent, an inching mechanism being employed.

(a)

(b)

Plate 31 Limestone feeder

A heavy apron feeder conveyor for limestone employing three strands of 200,000 lbf breaking load conveyor chain with steel slats bolted to integral K2 attachments situated on both sides of each chain at every pitch. The feeder is designed to handle up to 450 ton of material per hour at a speed of 15 ft/min.

Plate 32 Heavy duty inclined conveyor

This feeder handles rock. The conveyor must contend with high output and impact loads of considerable magnitude due to heavy rocks dropping on to the feeder from a considerable height. Heavy section steel slats are essential to resist deflection. Overlapping sections are necessary to prevent rock falling between the plates. The view shown is of the free hanging return section.

6

Scraper Conveyors

Scraper conveyors are of two general types, these being the scraper plate and the box scraper. Both are usually employed for handling bulk materials having free flowing characteristics.

6.1 Scraper plate conveyor

DESCRIPTION AND CHAIN TYPE

The scraper plate conveyor employs either a single strand or two strands of chain, to which are attached scraper plates or bars. The plates scrape along a continuous trough to convey the material; typical arrangements are shown in Fig 6.1. The bulk material is fed to the conveyor by a gravity chute arranged either at the side or through the top unloaded run. In either case, the aim should be to minimise direct contact of the load with the chain, this can be assisted by the use of deflector plates as shown.

The form of the material being conveyed depends on several variables, notably the material angle of repose, incline of conveyor, chain speed and effectiveness of material in-feed. The general form is typified in Plate 33 which shows a bagasse conveyor employed in the sugar cane industry.

In arriving at the proportions of this type of conveyor, the depth of material to be conveyed should not exceed two thirds of the trough depth. Chain speed is usually not more than 100 ft/min even when free flowing granular material is being handled. Lumpy material can be handled provided the linear spacing of the scraper plates is approximately three times that of the largest lump. The scrapers and trough may be of steel or wood, as determined by the class of material conveyed. Discharge takes place at the end of the trough or through apertures in the trough base; the latter may be gate operated, providing a controlled feed to specified positions. Various feeding and discharge arrangements are shown in Fig 6.2.

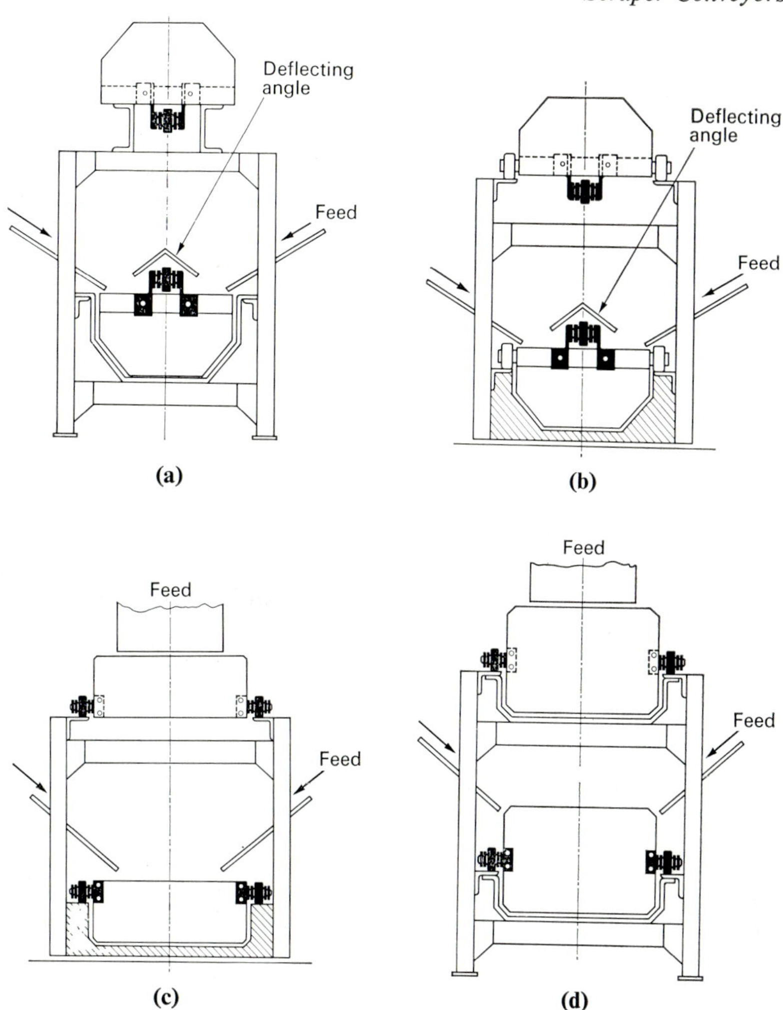

Fig 6.1 **(a)** Single strand of chain supported only by scraper plates. Suitable for light duty and short centre conveyors **(b)** Single strand system with outboard rollers mounted on the scraper plates **(c)** Two strand system conveying material on bottom run **(d)** Double depth flights attached to two strands of chain conveys material on top and bottom strand run.

Referring to Fig 6.1 the layout shown at (a) is suitable only for light duty applications at short centres, while those shown at (b), (c) and (d) are admirably suited to high capacity installations at long centre distances where several feed and discharge points are required. The chain rollers and outboard rollers minimise friction, thus reducing the overall power requirements. Inclined sections can be included, but these reduce the carrying capacity. In the case of a system combining horizontal and inclined runs, the capacity will be limited to that of the latter sections, the maximum permissible incline being about 35°.

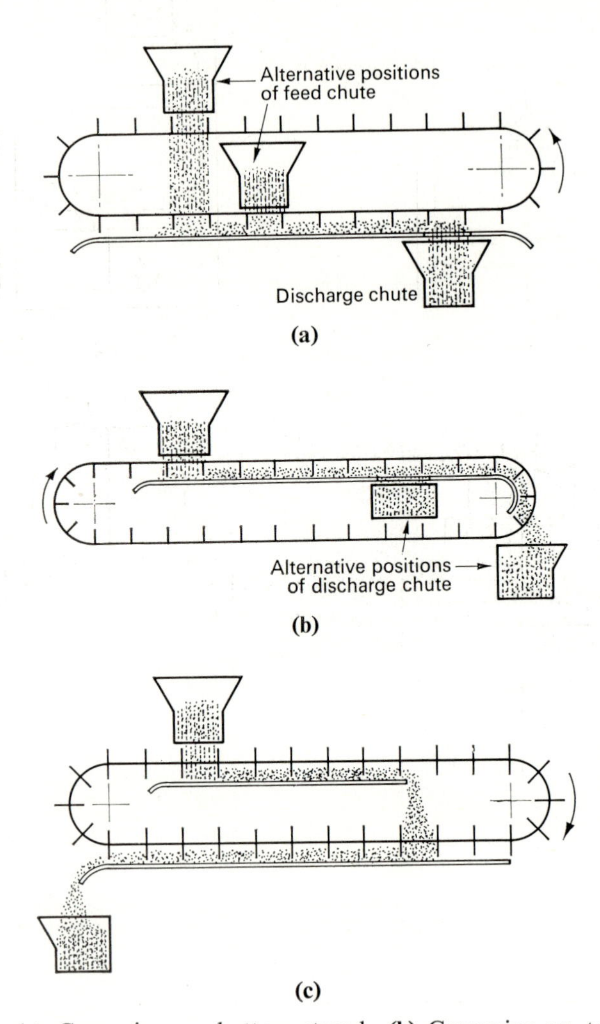

Fig 6.2 **(a)** Conveying on bottom strand **(b)** Conveying on top strand **(c)** For use with double depth flights where conveying is required on top and bottom strands.

Solid bearing pin chains are normally used due to the rough usage to which scraper conveyors can be put, scrapers being fitted to the chains by F or K attachments. Scrapers must be sufficiently strong to resist undue deflection which could cause 'bowing' of the pair of chains.

It is essential that the chains are kept in correct adjustment, since the cumulative weight of scrapers and chain can be considerable. If excess slack accumulates directly after the driver wheels, the free length of chain is liable to toggle and cause fouling at the lead-on to the support tracks.

The wheel size at the feed end is often governed by the method of feed, if a side feed is used then adequate space is necessary to accommodate the chute. Otherwise cast iron 8 tooth wheels should be adequate for most applications.

CHAIN PULL CALCULATIONS

The material feed and discharge points are generally located inboard of the driver and driven wheel centres. For clarity, however, it will be assumed that the material is being conveyed between points C and D in Fig 6.3.

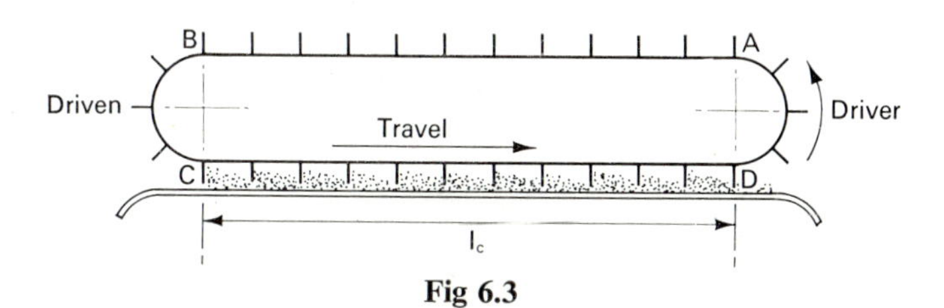

Fig 6.3

Total chain pull comprises the following component load pulls.

Section	*Load pull required*
A—B	To move the chains and scrapers on the unloaded side.
B—C	To turn the driven wheels.
C—D	To move the load.
	To move the chains and scrapers on the loaded side.

Load pull will be:—

Section	*Load pull required*		
A—B	To move the chain(s) and scrapers	$=\mu_R l_C w$	(1)
B—C	To turn the driven wheel(s)	$=0{\cdot}05\mu_R l_C w$	(2)
C—D	To move the load	$=\mu_{S2} W$	(3)
	To move the chain(s) and scrapers	$=\mu_R l_C w$	(4)

Therefore total chain load pull (lbf) = (1) + (2) + (3) + (4)
which simplified $= 2{\cdot}05\mu_R l_C w + \mu_{S2} W$

Where W = Weight of material on conveyor (lb)
w = Weight of chain(s) and conveying attachments (lb/ft)
l_C = Conveyor centre distance (ft)
μ_R = Overall coefficient of friction of chain rolling on track
μ_{S2} = Sliding friction between material and conveyor

$$\text{Running horsepower} = \frac{\text{Total chain pull (lbf)} \times \text{speed (ft/min)}}{33{,}000}$$

6.2 Box scraper conveyor

DESCRIPTION AND CHAIN TYPE

As in the case of the scraper blade conveyor, the box scraper type can use either a single strand or two strands of chain. The general construction is that of an enclosed box or trunking in which the chain is submerged in the material. The conveying movement relies on the 'en masse' principle, where the cohesion of the particles of material is greater than the frictional resistance of the material against the internal surfaces of the box. Because of this feature, a remarkably large volume of material can be moved by using flights of quite small depth.

The bottom surface of the box which supports the material, is also used to carry the chain (Fig 6.4).

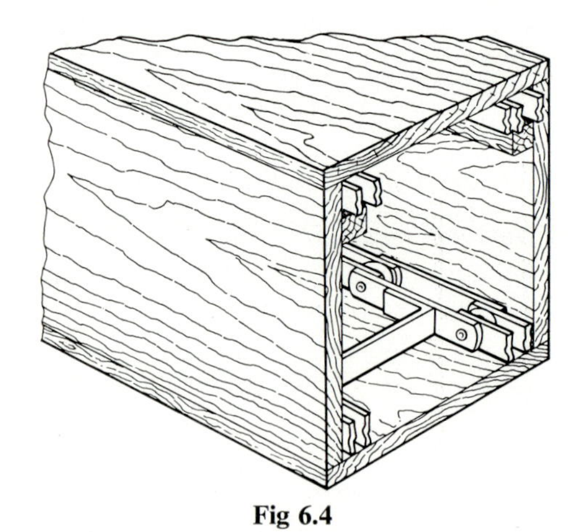

Fig 6.4

When conveying non-abrasive and free flowing material, such as grain, a chain speed of up to 100 ft/min is practicable. For aeratable materials however, as for example cement or pulverised starch, the chain speed must be reduced to 50 ft/min maximum. Excessive speeds reduce efficiency as the chain and attachments tend to pull through the material, leaving the top strata either stationary or moving at reduced speed; furthermore, turbulent conditions may be set up. For stringy, flaky and sticky materials, a speed of 40 ft/min should not be exceeded. Abrasive materials increase the amount of maintenance required, and to keep this within reasonable bounds the chain speed should not be more than 30 ft/min.

The operating principle depends as stated on the material having free flowing properties. Conveying can if desired be carried out contra-directionally by using both top and bottom runs of the chain. Operation is confined to straight sections, but these may incline from the horizontal. The amount of inclination is largely governed by the repose angle of the material and depth of the scraper. Effectiveness of flow can be prejudiced on inclines, and when handling grain for example, the maximum inclination from the horizontal should not exceed 15°.

The feed-in to the box may take the form of a manually fed side chute or a hopper with regulated feed. Alternatively the chain assembly itself can function as a regulator, as shown in Fig 6.5.

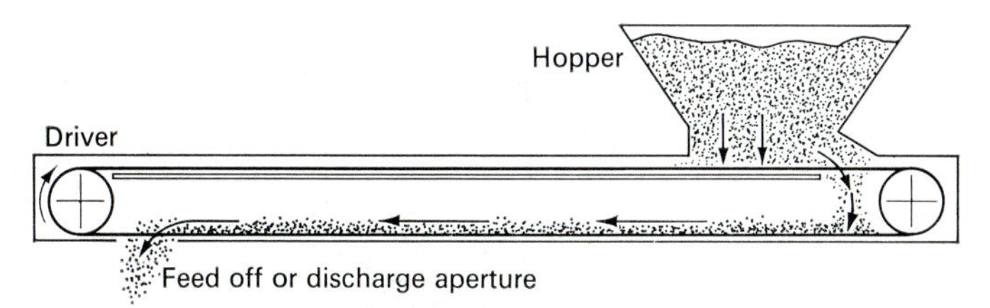

Fig 6.5

For scraper conveyors operating under uneven floor conditions, e.g. coal mines, it is common practice to drive from the opposite end to normal. This means that the drive is coincident with the on-loading position, the aim being that the conveying side at the top becomes the slack side and tends more readily to maintain its sliding position on the trough base. A typical installation is shown in Plate 37.

Standard chains of any size or strength may be adapted to the system. It is usual for the chain rollers and scraper plate depth to be equal to chain plate depth thus ensuring that as flat a surface as possible is moving along the scraper floor. Bush chains are sometimes used, but the roller type is more effective and is recommended. The chain pitch is normally governed by the required proportions of the box allied to the linear spacing of the scrapers. Wheel sizes are generally governed by box proportions, 8 or 12 teeth are commonly used. Head and tail wheels are normally cast iron with cast teeth and should have relieved form to reduce packing of material between the chain and wheel teeth.

Integral L attachments as illustrated in Fig 6.6 may be used in either the single or double strand light duty systems.

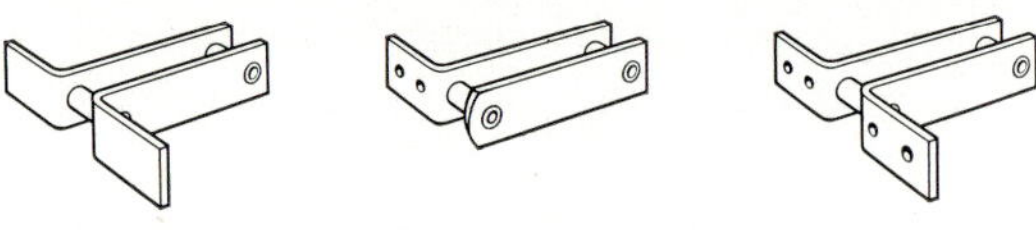

Fig 6.6

In addition to using the above general purpose L attachments special scraper attachments are available, an example of these being shown in Fig 6.7.

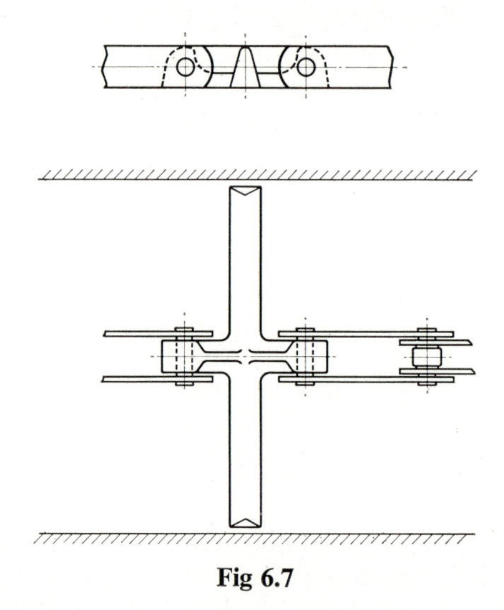

Fig 6.7

The advantages of this type of scraper are:—

Use of precision conveyor chain retains advantage of large bearing areas and high hardness at articulation points.

Weight of malleable scrapers tends to maintain the chain on the box base. This is further assisted by the angled face of the scraper.

Relatively high tensile working loads enable long centre distance conveyors to be used.

A continuous unbroken scraping surface is presented to the material being conveyed.

The limited area of the top of the malleable scraper reduces material carry-over across the discharge aperture to a minimum.

Whilst this type of chain has been introduced primarily for the handling of damp grain, as typified by Plate 38, it is used successfully for other free flowing granular materials such as animal food pellets, cotton seed, pulverised coal, cement, etc. The chain can operate in either direction but conveying of material can only be done on one strand of the chain, which in Fig 6.8 is the bottom strand.

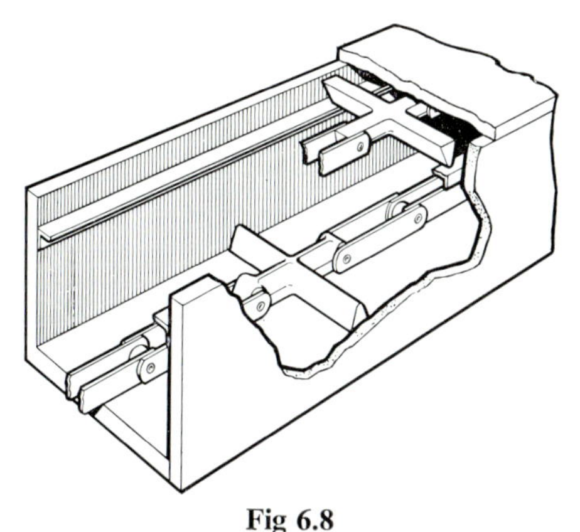

Fig 6.8

CHAIN PULL CALCULATIONS

The formula is given in Section 6.1.

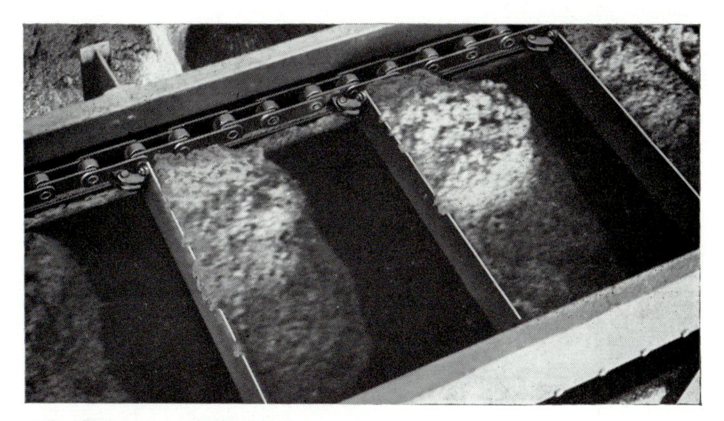

Plate 33 Bagasse scraper conveyor

In the sugar cane industry, cane is passed through cutting knives and grinding rolls until all the juice has been extracted. The pulped mass which is left is called 'bagasse' and is conveyed by means of scraper conveyors to the furnaces for use as fuel. These conveyors are often up to 300 ft long and to compensate for conveyor malalignment and uneven chain wear, hinged flights are used.

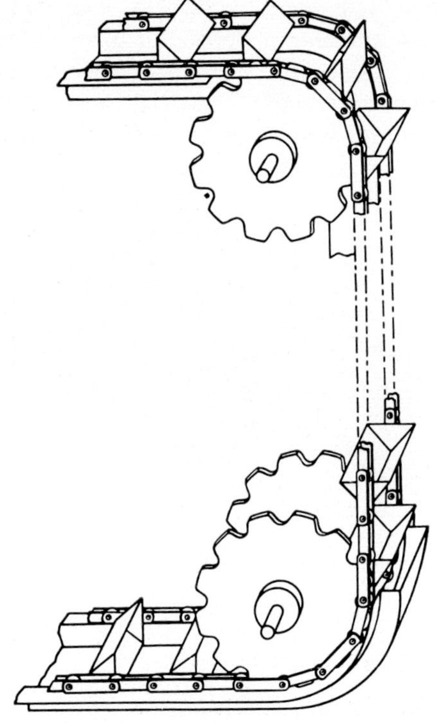

Plate 34 Combined scraper/elevator for coal and similar materials

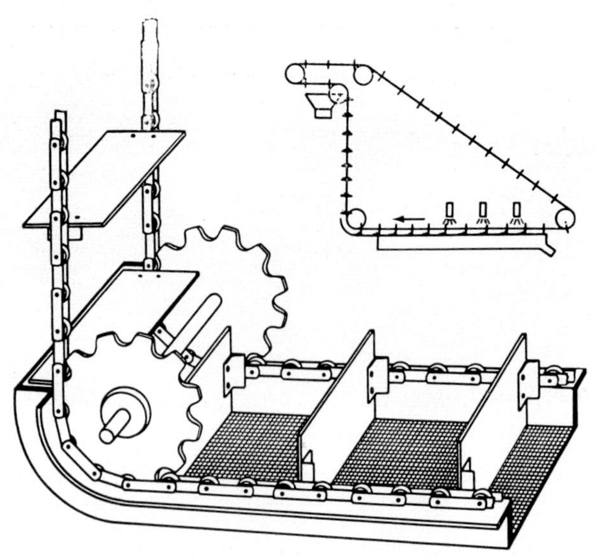

Plate 35 Clearing the waste from the surface of filters

Plate 36 Standard chain with scraper attachments riveted to chain plates—side feed chute is illustrated in the foreground

Plate 37 Underground coal scraper

For scraper conveyors operating under conditions where floor undulations and 'snaking' are unavoidable it is common practice to drive from the opposite end to normal. This means that the drive is coincident with the on-loading position, the aim being that the conveying side at the top becomes the slack side and tends more readily to maintain its sliding position on the trough base.

Typical examples are run-of-mine coal scraper conveyors operating at the coal face under long-wall mining methods.

(a)

(b)

Plate 38 Scraper conveyor handling grain

Plate **(a)** shows the drive end of the conveyor and the method of supporting the return strand of chain. The feed chute shown in **(b)** illustrates the method of feeding grain through the return strand of chain.

Plate 39 Fish offal inclined scraper conveyor

Fish offal being conveyed to a fish meal plant. Here the chains must contend with the corrosive effects of water and fish oils. Some protection for the chains is afforded by running them inboard of the guide tracks. On the loaded strand the chains run on their rollers, return strand support is obtained by bearing the tops of the scrapers on guide tracks. Standard chains in normal materials were used on grounds of economy, but special materials would increase chain life. Regular cleaning is an essential part of maintenance on this type of conveyor.

Plate 40 Coal scraper conveyor

This conveyor uses steel plates bolted between a pair of chains by F attachments. The return strand runs over the working strand to give a free discharge below the trough. Levers and quadrants for selection of discharge hatch are to be seen on the side of the trough. In view of the dirty operating conditions larger chain rollers than standard are fitted to ensure rotation and prevent skidding. Whilst this application concerns the handling of coal, the design and principles involved are equally applicable to the handling of any reasonably dry and free flowing bulk material both in the horizontal plane and on inclines.

Plate 41 Sludge scraper conveyor

Sludge in the form of fine metal swarf in cutting oil is deposited in a settling tank and then removed by inclined scraper conveyor. Corrosive and abrasive conditions dictate a chain with a high safety factor in the interests of obtaining good wearing areas. Given this factor, standard chains are satisfactory. On the scraping side of the conveyor, the scraper edges run on the trough base with the chains taking up a catenary between. The central projecting member fixed to each scraper is to obviate toggling when scraping.

Plate 42 Elevator from quench tank

Small components are transferred to further operations from a quench tank after heat treatment by means of this plate scraper elevator. In this application the return strand is carried over the working strand as shown in the photograph. Three anti-toggling vanes are fitted behind each blade. Chains on both working and return strands roll on guide tracks, the chains being mounted outboard of the scraper blades for this purpose.

7

Pallet Conveyors

Description and chain type 7.1

The pallet conveyor comprises a single strand of chain positioned on its side with the bearing pins vertical. The chain operates in a continuous closed circuit and is often referred to as a Carousel conveyor. Slats are fixed to the chain to form a continuous carrying platform, but as the slats are supported independently, the chain functions as a pulling medium only as shown in Fig 7.1.

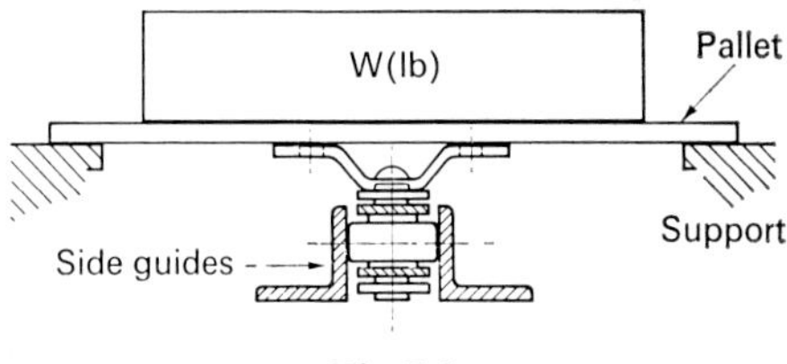

Fig 7.1

This type of conveyor can be found in virtually all industries, and because of its inherent flexibility the system can be arranged to complement production processes. The continuous circuit feature for example enables workboxes, cartons etc., to be returned to their initial stations. Large radiused bends are frequently incorporated in circuits using this system, and consequently the chain uses tracked bends to guide the chain round the circuit. For lateral stability side guide tracks should be introduced over the straight sections of the circuit in addition to those controlling the chain path around bend sections. Lengths of circuits are virtually unrestricted since multiple drive points can be introduced into any one circuit. These drives normally being of the caterpillar type.

The unsupported chain, in the vertical plane between the points of attachment to the slats, should be restricted to a length of about 5 ft. Furthermore where caterpillar drives are used recessed guide tracks positioned below the main chain are necessary to control the sag of the chain to ensure correct gearing. The length of tracking should extend over the caterpillar drive distance by 4 pitches at each end; reference should also be made to Chapter 15 in this connection.

Actual slat support is normally effected by one of the following three methods.

Sliding the underside of the slats on suitable guide tracks. If the slats are of wood it is common practice to attach brass wear pads to the slats.

Providing rollers on the ends of each slat, these rollers running on guide tracks.

Running the slat on idler rollers as in Fig 7.2.

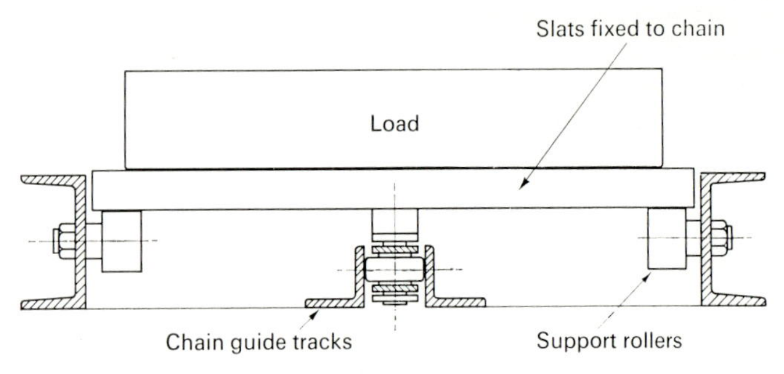

Fig 7.2

Hollow or solid bearing pin chain can be employed, chain pitch being largely dictated by the size of pallet. Maximum pitch consistent with good running performance should be aimed at in order to reduce cost. Coned bearing pins (see Chapter 9) are used on installations with slight changes in elevation. At tracked bend sections high roller loadings can be induced, therefore roller material should be selected to deal with the resulting pressure as outlined in Chapter 2.

The provision of adjustment to cater for chain elongation is essential. In circuits where tracked bends are used then adjustment is achieved by movement of the bend section.

On installations where wheels are used at bends, sizes are generally dictated by the circuit path to be followed and the size of the pallets. The wheel boss should allow adequate clearance for free passage of the pallets.

Pallets can be fixed to the chain by many methods. The usual method of fixing the pallet to the chain is by G attachments or connecting links with extended bearing pins. Arrangements making use of hollow bearing pin chain, or holes in the outer link plates, are also used.

Chain pull calculations 7.2

It is assumed that the chain pallets will be supported on rollers as shown in Fig 7.2, with the conveyor circuit following a path as indicated in Fig 7.3 and driven by a caterpillar drive of the type described in Chapter 15.

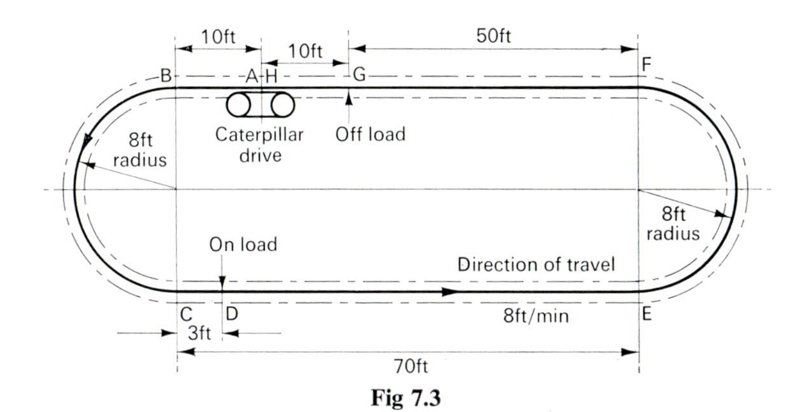

Fig 7.3

Referring to Fig 7.3
Total chain pull comprises the component load pulls:—

Section	*Load pull required*
A—B	To move the chain, slats and fixing media.
B—C	To overcome the reaction load of the chain with its rollers bearing on the curved guide track. To move the chain, slats and fixing media around the curved section.
C—D	To move the chain, slats and fixing media.
D—E	To move the dead load, chain, slats and fixing media.
E—F	To overcome the reaction load of the chain with its rollers bearing on the curved guide tracks. To move the dead load, chain, slats and fixing media round the curved section.
F—G	To move the dead load, chain, slats and fixing media.
G—H	To move the chain, slats and fixing media.

EXAMPLE

Where W = Weight of material on conveyor = 9,940 lb
w_M = Weight of material on conveyor = 70 lb/ft
w = Weight of chain and conveying attachments = 7 lb/ft
w_S = Weight of slats and fixing bolts = 50 lb/ft
μ = Overall chain coefficient of friction = 0·18

μ_{R2} = Overall coefficient of friction of supporting rollers on track = 0·025

l_S = Length of various sections where applicable

r_B = Radius of bend = 8 ft

e = 2·718

Section	*Load pull required*	
A—B	$l_S(w+w_S)\mu_{R2}=10(7+50)\times 0{\cdot}025$	$=14{\cdot}2$
B—C	$l_S(w+w_S)\mu_{R2}e^{\mu\theta}=14{\cdot}2\times 2{\cdot}718^{\,0{\cdot}18\pi}$	$=25$
	$\pi r_B(w+w_S)\mu_{R2}=8\pi(7+50)\times 0{\cdot}025$	$=35{\cdot}8$
C—D	$l_S(w+w_S)\mu_{R2}=3(7+50)\times 0{\cdot}025$	$=4{\cdot}3$
D—E	$l_S(w_M+w+w_S)\mu_{R2}=67(70+7+50)\times 0{\cdot}025$	$=212{\cdot}7$
	Part total chain pull	$=292$
E—F	Part total chain pull $\times e^{\mu\theta}=292\times 2{\cdot}718^{\,0{\cdot}18\pi}$	$=514$
	$\pi r_B(w_M+w+w_S)\mu_{R2}=8\pi(70+7+50)\times 0{\cdot}025$	$=79{\cdot}8$
F—G	$l_S(w_M+w+w_S)\mu_R=50(70+7+50)\times 0{\cdot}025$	$=158{\cdot}8$
G—H	$l_S(w+w_S)\mu_R=10(7+50)\times 0{\cdot}025$	$=14{\cdot}2$
	Total chain pull	$=766{\cdot}8$ (lbf)

$$\text{Horsepower}=\frac{\text{Total chain pull(lbf)}\times\text{speed (ft/min)}}{33{,}000}$$

$$=\frac{766{\cdot}8\times 8}{33{,}000}=0{\cdot}186$$

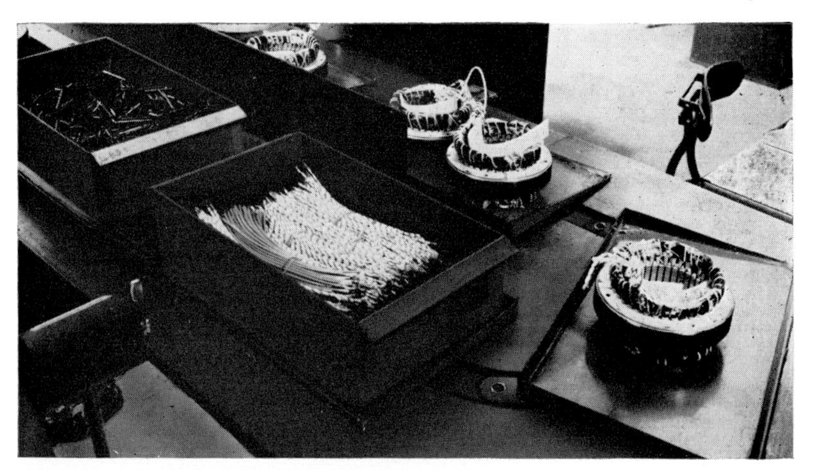

Plate 43 Light assembly pallet conveyor

In this conveyor, the chain is accommodated in the top of the assembly bench and tray shaped pallets are secured by bolting through the hollow bearing pins of the chain by readily detachable countersunk bolts. The trays slide on the flat metal top of the bench. Since the bend radii are small, normal toothed wheels are employed at these points, one of the wheels being driven. Both strands of the system are usefully employed by making the assembly bench double sided.

Plate 44 Newspaper parcelling conveyor

This conveyor is fitted in a newspaper despatch department. Wood slats are employed, but instead of being fitted across a pair of chains as in slat conveyors, they are mounted on G attachments on one chain to articulate in the horizontal plane. The radius of curvature is large and, therefore, the straight slats can articulate with only a small clearance between each. In this installation the slat ends slide on support angles. Drive is by caterpillar drive chain with driving dog teeth, and this feature is clearly illustrated in the foreground.

Plate 45 Foundry mould pallet conveyor

This type of conveyor is favoured in mechanised foundries to convey moulding boxes from moulding machines through the cycle of pouring, cooling, knock-out, sand removal and return of empty boxes to moulding machines. The pallets have convex leading and concave trailing edges for bend articulation and are fixed rigidly to the chain by G attachments. Pallets are supported on static rollers in the conveyor bed. Side guides, bearing on the chain rollers, control the path of the chain on bends and on straight runs. The drive is by caterpillar chain. Conditions are severe due to abrasion and a high selection factor is allowed when selecting the chain. A dry lubricant such as colloidal graphite is preferable.

8

Haulage

Traverser chains 8.1

DESCRIPTION AND CHAIN TYPE

In these systems the chain is employed purely as a pulling medium. The simplicity of the basic principle is shown in Fig 8.1 where it will be seen that a wheeled bogie is permanently attached to the chain.

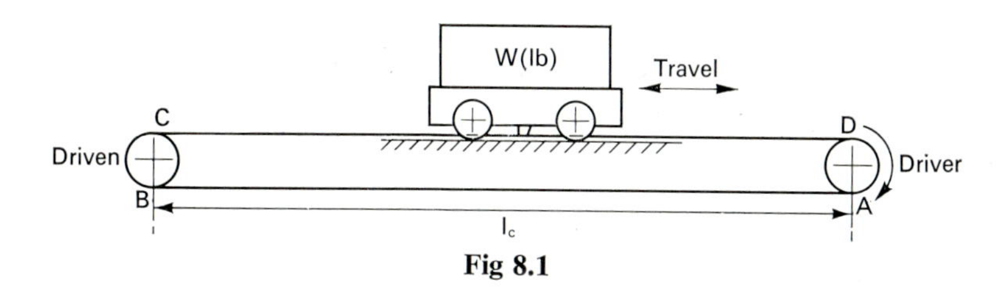

Fig 8.1

In all cases, it is desirable to use a single chain to eliminate any possibility of unequal load distribution over a number of chains. However, a series of chains may be essential for handling large bogies, as exemplified by the installation shown in Plate 48. Attachment methods vary according to the type of installation, but basically the chain may have its two ends fixed to the bogie, or be endless with a suitable attachment point. The chain must be of solid bearing pin type, with the largest pitch which is practicable, bearing in mind the requirement of an adequate number of wheel teeth. If chain speed is slow or operation infrequent then bush chain can be used.

Maintenance of correct chain tension is essential, particularly in installations where there is reversal under full load with the drive at one end of the system. The usual adjustment method is by outward movement of the tail wheel, but an alternative arrangement when a non-endless chain is used is to introduce a turnbuckle device at the attachment point of the chain to the bogie.

In designing the drive it is usually beneficial to include a fluid coupling or similar device to provide a smooth take-up. When several chains are mounted on wheels sharing a common drive shaft, it is essential that the shaft is so proportioned as to reduce torsional deflection to a minimum to prevent unequal load distribution on the chains. Wheels having a minimum of 12 teeth are recommended and where the installation is subjected to frequent starting-up under load, they should be manufactured from steel.

8.2 Chain pull calculations

When the chain is operating under impulsive conditions with frequent starting and stopping of the load, both the start-up friction and accelerating force must be taken into account; the latter being required to overcome the total mass of load plus bogie, and the inertia of rotating components such as axles and wheels.

The equivalent mass m to be accelerated is:—

$$\left(\frac{W+W_G}{g}\right)+\frac{2W_W}{g}\left(\frac{k_W}{r_W}\right)^2 \qquad (1)$$

Where W = Weight of load (lb)
W_G = Weight of bogie (lb) (including wheel and axle weights)
W_W = Weight of each axle plus its pair of wheels (lb)
k_W = Radius of gyration of wheels (in)
r_W = Radius of wheels (in)
g = Acceleration due to gravity = 32·2 ft/sec²

The accelerating force is a function of the time taken to attain maximum speed from rest and the general formula:—

$$F_A = mf \text{ (lbf)} \qquad (2)$$

can be applied where

F_A = Accelerating force (lbf)
m = Mass to be accelerated
f = Acceleration rate (ft/sec²)

$$= \frac{\text{Maximum speed (ft/sec)}}{\text{Time to attain max. speed (sec)}} = \frac{v_{max}}{t_{max}}$$

Substituting expression 1 into formula 2

$$F_A = \frac{\left[(W+W_G)+2W_W\left(\frac{k_W}{r_W}\right)^2\right]}{32{\cdot}2} \times \frac{v_{max}}{t_{max}} \text{ lbf} \qquad (3)$$

Referring once again to start-up friction, this varies with each set of specific conditions but it is acceptable in a theoretical assessment to take running friction and increase this by an appropriate factor. The force to overcome starting friction is then:—

$$F_F=(W+W_G)\,\mu_{R6}\,K_W \qquad (4)$$

Where μ_{R6} = Overall coefficient of friction of the bogie wheels on track
K_W = Constant, taken as 1·10 for wheels with roller bearings and 1·50 for plain bearings

In many cases the weight of bogie wheels and axles is small compared to the combined weights of bogie and dead load and can be ignored. Formula (3) would then become:—

$$F_A=\left(\frac{W+W_G}{32\cdot 2}\right)\times\frac{v_{max}}{t_{max}}\ \text{lbf}$$

So far the weight of the chain has not been taken into account. This should be included in formula 3 and allowance made in formula 4. The total accelerating force will now be:—

$$F_A=\frac{\left[(W+W_G)+2W_W\left(\frac{k_W}{r_W}\right)^2+wl_C\right]}{32\cdot 2}\times\frac{v_{max}}{t_{max}}\ \text{lbf} \qquad (5)$$

Similarly F_F will now be

$$F_F=(W+W_G)\mu_{R6}K_W+2{\cdot}05\ l_C\mu_{R6}w \qquad (6)$$

Where w = Weight of chains and attachments (lb/ft)
l_C = Conveyor centres (ft)

Total force (F_T) to move the dead load, bogie and chain from rest to maximum speed

$$F_T=F_A+F_F\ \text{(lbf)}$$

In cases where the rate of acceleration is low and continuous smooth running predominates, it is necessary only to consider the sustained pull required to maintain the loads in motion. In such conditions, the assessment would follow normal procedure. Referring again to Fig 8·1, the load pull is as follows:—

Section	*Load pull required*	
A—B	$wl_C\mu_{R6}$	(7)
B—C	$0{\cdot}05\ wl_C\,\mu_{R6}$	(8)
C—D	$wl_C\,\mu_{R6}$	(9)
	$(W+W_G)\mu_{R6}$	(10)

Total load pull = (7)+(8)+(9)+(10) which simplified
$= (W+W_G)\mu_{R6}+2{\cdot}05\ wl_C\,\mu_{R6}$ (lbf)

If the tractive effort to move the bogie is known then this would be expressed in lbf effort per ton from which the μ_{R6} value could be obtained. Alternatively an assessment of μ_{R6} is possible, in line with Chapter 2 if the bogie wheel proportions and bearings are known.

Plate 46 Molten steel ladle conveyor

In this foundry ladle repair shop, the bogie is hauled by an underfloor conveyor chain, running over head and tail wheels and carrying a towing attachment, coupled to the bogie via a slot in the floor. Since load movement occurs in both directions, it is essential that the chain is adjusted with minimum slack to avoid 'jumping' the wheels. In view of the low speed, the chain rollers were dispensed with and a 'bush' chain fitted.

Plate 47 Underground mine tub conveyor

This system uses standard conveyor chain with coned bearing pins which allow a radius of 8 feet to be negotiated. The method was evolved to reduce the underground excavation necessary to contain the system, small pits at only head and tail ends were required. Working and return strands are in the same plane so that, if required, both strands could be used for load propulsion.

Plate 48 Trawler slipway haulage *(overleaf)*

To haul trawler bogies weighing 200 ton, four strands of 85,000 lbf breaking load chain are used. The following special considerations apply to this type of duty.
— The chains must be liberally coated with water repellent grease to combat sea-water corrosion.
— Dimensions D_1, D_2, D_3 will not necessarily be equal but will be determined by reference to the weight distribution of the vessels handled to load each chain equally.
— The long headshaft must be of suitable material and proportions to minimise twist, and the prime mover should be situated at mid-point or thereabouts, for the same reason.
— Chain connection to the bogies can be by normal chain connecting links, but at the headshaft side turnbuckles should be introduced to compensate for possible uneven chain wear due to varying loading. For the same reason, each tail wheel should be independently adjustable.
In this instance, large rollers are fitted to the chain to facilitate passage of the chain return strand through a 9 in pipe.

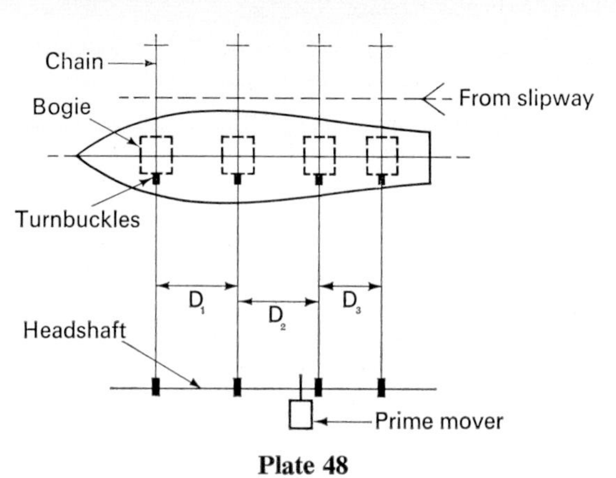
Chain
From slipway
Bogie
Turnbuckles
D1
D2
D3
Headshaft
Prime mover

Plate 48

Underfloor tow conveyor

DESCRIPTION AND CHAIN TYPE

In principle a series of small trolleys, or plain slipper pads, are permanently connected at intervals to a single endless chain running on its side in the horizontal plane as shown in Figs 8.2 and 8.3. The trolleys are pulled around the circuit by the chain and at the same time provide support for the chain. Trucks, or similar, are engaged and disengaged, with the trolleys at any desired position in the circuit by the use of a drop bar connection. The trucks thus follow the circuit path of the chain.

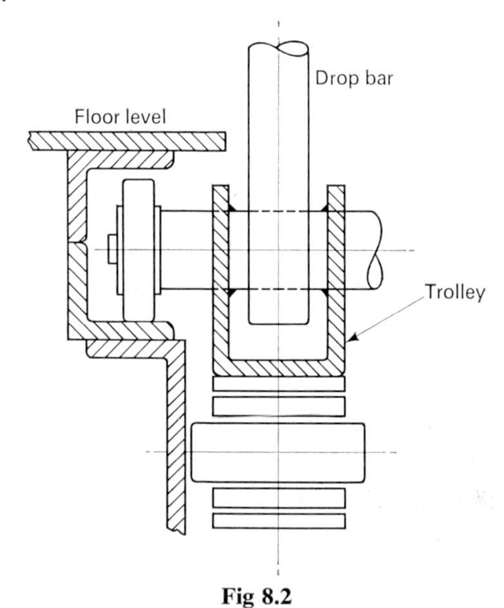

Fig 8.2

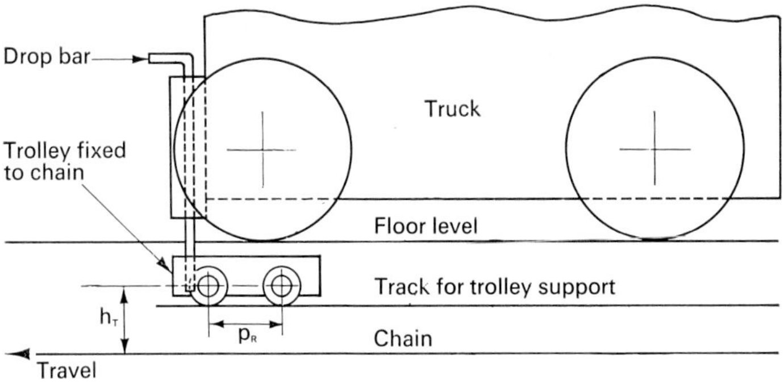

Drop bar
Truck
Trolley fixed to chain
Floor level
Track for trolley support
h_T
p_R
Chain
Travel

Fig 8.3

This type of conveyor system is normally restricted to factory locations where its flexibility makes it admirably suitable for the linking of machines or feeding components from storage to assembly or processing points. Complex circuit paths can be negotiated and the overall lengths are virtually unrestricted since multiple drives can be used.

Guiding in the horizontal plane is possible by bearing the chain rollers on side guide tracks and large bend radii can be negotiated. These may be desirable for particular layout conditions and to suit the turning or 'castoring' characteristics of the trucks.

At tracked bend sections high roller loadings can be induced and therefore roller material should be selected to cope with the resulting pressures. If large radiused bends are employed then toothed wheels cannot always be used for driving. It then becomes necessary to use a caterpillar drive. For lateral stability chain guide tracks should be employed on both straight and curved sections of the circuit.

The trolleys or pads may be fixed to the chain by hollow bearing pins, drilled link plates or extended bearing pins. There is a limit to the length of chain which can be freely suspended between trolley positions. This is normally limited to 5 ft so that if propelling trolleys were required at say 10 ft then an intermediate trolley would be required solely to provide adequate chain support.

The provision of adjustment to cater for chain elongation is essential. In circuits where it is not possible to use toothed wheels with 180° lap then the only effective method is by the outward movement of a complete track bend. Where chain wheels are employed then the number of teeth is dictated by the circuit path to be followed and the 'castoring' effect of the trucks.

CHAIN PULL CALCULATIONS

For illustration the following circuit path and applicational details are assumed.

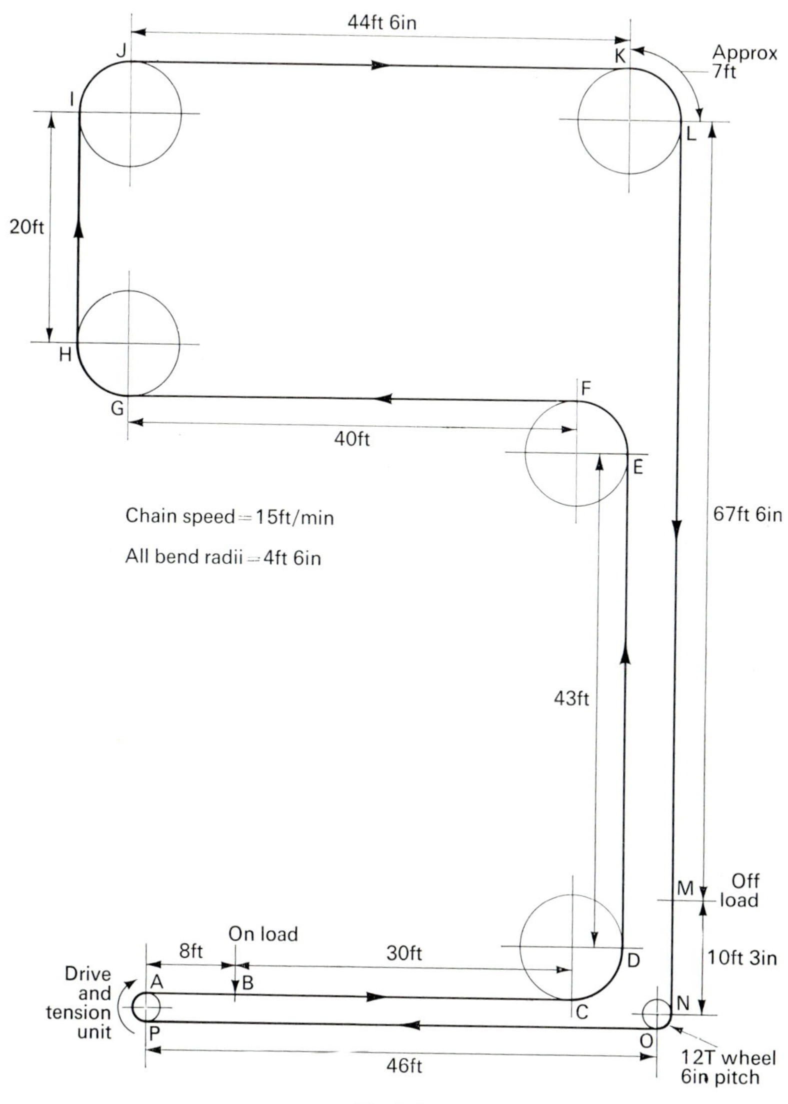

Fig 8.4

W_T = Weight of each truck, fully loaded = 560 lb
w = Weight of chain and attachments = 8 lb/ft
l_T = Spacing of trucks = 10 ft
p_R = Pitch of trolley rollers = 6 in
h_T = Distance from trolley roller centre to centre line of chain = 2·625 in
μ_R = Overall coefficient of friction of chain rolling on track = 0·15
μ_{R4} = Overall coefficient of trolley roller on track = 0·13
μ_{R5} = Overall coefficient of truck roller on track = 0·021

It is simpler when dealing with complex circuits to use the section method for calculating the chain pull as account need only be taken of the actual length per section, due allowance being also made for the effect of reaction loadings induced at bend sections and at idler wheel positions. Total chain pull comprises the component load pulls:—

1. To move the truck against the truck roller friction

$$\frac{W_T \mu_{R5}}{l_T} = \frac{560}{10} \times 0{\cdot}021$$
$$= 1{\cdot}18 \text{ lbf/ft}$$

2. To move the chain and trolleys against the trolley roller friction

$$= w\mu_{R4}$$
$$= 8 \times 0{\cdot}13$$
$$= 1{\cdot}04 \text{ lbf/ft}$$

3. To overcome the reaction load at the trolley

$$= \frac{W_T\ \mu_{R5}\ h_T\ \mu_{R4}}{l_T\ p_R}$$

$$= \frac{560 \times 0{\cdot}021 \times 2{\cdot}625 \times 0{\cdot}13}{10 \times 6} = 0{\cdot}07 \text{ lbf/ft}$$

Tracked bend factor $= e^{\mu_R \theta}$
$= 2{\cdot}718^{\,0{\cdot}1 \times 1{\cdot}57}$
$= 1{\cdot}266$

Section	Length (ft)	Bend	Factor	Pull (lbf)	Total pull (lbf)
AB	8		1·04	8	8
BC	30		2·29	69	77
CD	7	 90°H	2·29 1·266	16	93 118
DE	43		2·29	98	216
EF	7	 90°H	2·29 1·266	16	232 294
FG	40		2·29	92	386
GH	7	 90°H	2·29 1·266	16	402 509
HI	20		2·29	46	555
IJ	7	 90°H	2·29 1·266	16	571 723
JK	44·5		2·29	102	825
KL	7	 90°H	2·29 1·266	16	841 1,065
LM	67·5		2·29	155	1,220
MN	10·25		1·04	10	1,230
NO	1·5	90°H Wheel	1·04 1·03	2	1,232 1,269
OP	46		1·04	48	1,317

Total chain pull = 1,317 lbf

$$\text{Running horsepower} = \frac{1{,}317 \times 15}{33{,}000} = 0{\cdot}6$$

Plate 49 Underfloor tow conveyor handling work trucks

The chain is suspended from and tows trolleys running on underfloor tracks. The drop bolts at the front which engage via the floor slot, can be withdrawn automatically by a front plough, should an obstruction be encountered.
The chain is of the hollow bearing pin type with bolts passing through the hollow bearing pins for attachment to the trolleys. Drive is by side mounted caterpillar drive of the hinged dog type.

9

Conveying Directly on Chain Plates or on Rollers

In these conveying systems, two or more chains are arranged parallel to each other. The load is carried directly on the chain rollers or side plates with the chain also being the traction medium.

Conveying directly on chain plates 9.1

DESCRIPTION AND CHAIN TYPE

In this system, the load is carried on the chain plates, the chain presenting a continuous surface to the underside of the load. This type of conveyor is intended for handling rigid unit loads, which in some applications can be halted whilst the chains remain in motion. Two types of chain are normally used: deep link plate and block chain. With the deep link plate chain the loaded strand of the chain is usually supported on rolled steel angles, the rollers protruding beyond the link plates underneath. On the return strand, raised tracks are necessary in order to contact the rollers, or alternatively the link plates may slide on the tracks, or be supported by plain idler rollers.

Chains are also available with coned bearing pins, this enables a limited amount of curvature to be accepted in what is normally the non-articulating plane. The radius of curvature depends on the pitch and degree of coning that can be applied successfully. Large chain pitches must be avoided, as the reaction loads at bends can cause unacceptable bending stresses in the chain plates. On the curved sections, the chain should be guided by tracks made of phosphor bronze or mild steel, the tracks being recessed to clear the ends of the bearing pins. Cast iron wheels having a minimum of 8 teeth with cast tooth form, are suitable for the majority of deep link plate applications.

Fig 9.1 illustrates a block chain used principally for conveying crates in dairies and breweries etc. With this chain the configuration of the outer plates protect and shield the bearing pin ends. The shaped inner link and coned bearing pins allow the chain to negotiate relatively small horizontal bends. Chain plate edges are hardened to resist abrasive wear. On the loaded run the chain is supported in a rolled steel channel which is formed to suit the desired circuit layout. The unloaded strands are normally supported on plain faced idler rollers or alternatively channel tracks may be employed.

Wheels for this type of chain have every alternate tooth removed to clear the solid inner link. Steel fabricated wheels having a minimum of 12 theoretical teeth, i.e. 6 actual teeth are suitable for the majority of applications.

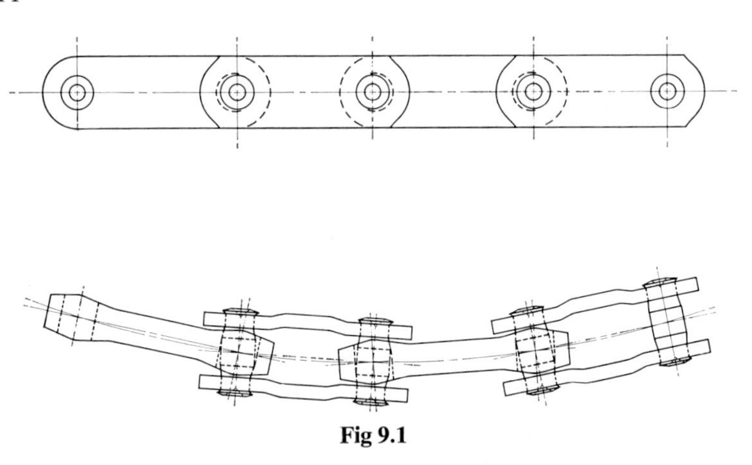

Fig 9.1

With both types of chain, from the point where the chain leaves a horizontal bend section then a straight section of 4 to 5 chain pitches must be introduced before the chain engages with a chain wheel. This will counteract any crabbing action which the chain may have acquired when traversing the bend. In applications using coned pin chains it is possible to have two or more chains of unequal length, due to each chain moving along a different curvature, and in such cases an independent adjustment point for each tail wheel is necessary.

CHAIN PULL CALCULATIONS

With the load resting on the plate edges and moving at the same speed as the chain as Fig 9.2 the total chain pull comprises the following component pulls, where:—

W = Weight of material on conveyor (lb)
w = Weight of chain(s) and conveying attachments (lb/ft)
l_C = Conveyor centres (ft)
μ_R = Overall coefficient of friction of chain rolling on track

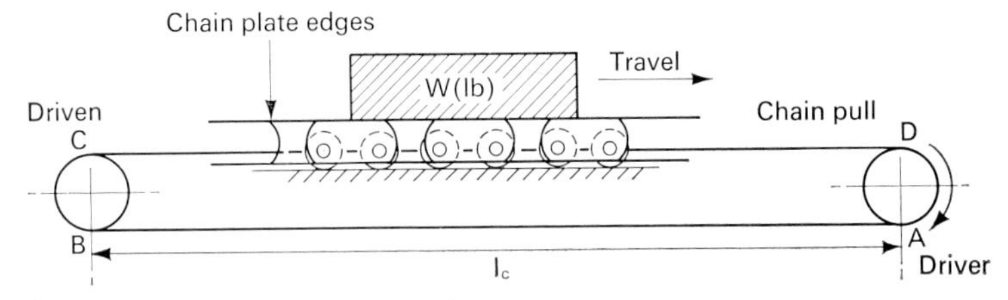

Fig 9.2

Section	*Load pull required*	
A—B	To move the chains	$=\mu_R l_C w$
B—C	To turn the driven wheels	$=0{\cdot}05\mu_R l_C w$
C—D	To move the dead load	$=\mu_R W$
	To move the chains	$=\mu_R l_C w$
	Total chain pull	$=2{\cdot}05\ \mu_R\ l_C\ w+\mu_R W$ (lbf)

If the chains were returned by sliding with their plate edges on guide tracks then the total chain pull would be:—

$$=l_C w(1{\cdot}05\ \mu_R+\mu_S)+\mu_R W \quad \text{(lbf)}$$

Where μ_S=Coefficient of friction of chain sliding on track.

With chains running and the load halted there will be an additional pull required to overcome the frictional resistance between the underside of the load and the chain plate edges.

If μ_{S6}=Coefficient of sliding friction between chain and load.

Total chain pull (assuming chains rolling on return side)

$$=2{\cdot}05\ \mu_R l_C w+W(\mu+\mu_{S6})$$

For coned bearing pin application the chain pull is calculated as for the plain pin but the additional load involved by guiding the chain around the horizontal bend sections must be taken into account as follows.

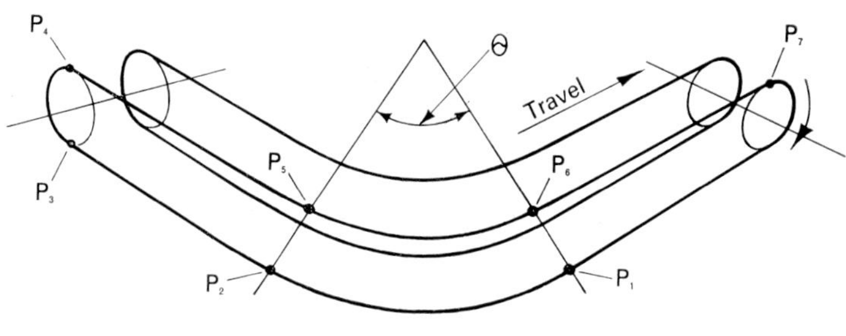

Fig 9.3

If the simple circuit shown in Fig 9.3 is considered, the value of load pull at the various positions can be identified by the following references.

Where θ = Bend angle (radians)

μ_S = Coefficient of sliding friction between chain and track.

Unloaded strands

P_1 = Chain pull at entry into bend section

P_2 = Chain pull at exit from bend section

$= P_1 \times e^{\mu_S \theta}$

P_3 = Chain pull at entry onto tail wheels

$= P_2$ + chain pull between position P_2 and P_3

Loaded strands

$P_4 = P_3 \times 1{\cdot}05$

$P_5 = P_4$ + chain pull between positions P_4 and P_5

$P_6 = P_5 \times e^{\mu_S \theta}$

$P_7 = P_6$ + chain pull between positions P_6 and P_7

= Total chain pull (lbf)

From this it will be appreciated that the introduction of bends has a cumulative effect on the load pull in the circuit.

9.2 Conveying directly on chain rollers

DESCRIPTION AND CHAIN TYPE

The use of load carrying rollers is limited to handling of rigid unit loads under clean conditions. No attachments are required, the load simply running on the chain rollers. It should be noted that the load will travel at twice the chain speed.

The chain may be of either hollow or solid bearing pin type. Chain pitch on complex and highly loaded systems is frequently dictated by the limitation of both individual roller loading under direct compression, and the pressure on the track at the point of roller contact. When high individual roller loadings are involved, mild steel case-hardened rollers will be necessary. Tracks using commercial rolled sections are suitable for simple installations but where heavy unit loadings are involved hardened tracks may be necessary. Tracks must have the running surface maintained in a clean condition at all times.

Wheels made of cast iron, with a minimum of 8 teeth of cast tooth form, are suitable for simple layouts.

CHAIN PULL CALCULATIONS

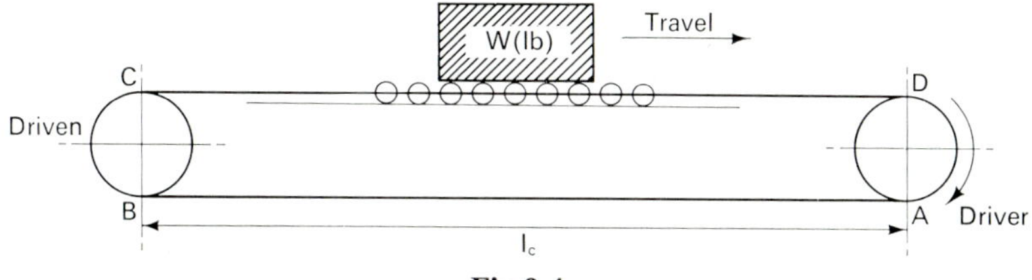

Fig 9.4

Referring to Fig 9.4 the total chain load pull is determined as follows:—

Section	*Load pull required*
A—B	To move the chains on the unloaded strands.
B—C	To turn the driven wheels.
C—D	To move the chains on the loaded strands.
	To move the dead load.
	To overcome the impingement resistance of chain rollers against the bushes.

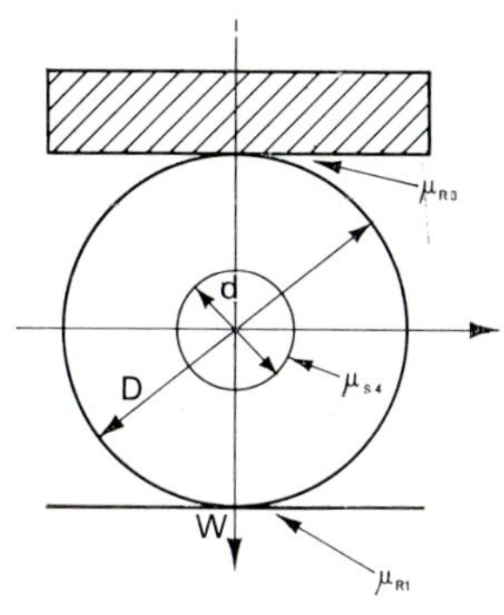

Fig 9.5

To review these loadings only one chain roller may be considered as Fig 9·5 and where:—

W = Weight of material on conveyor (lb)
w = Weight of chain(s) and attachments (lb/ft)
l_C = Conveyor centres (ft)
μ_{R1} = Coefficient of rolling friction between chain roller and track
μ_{R3} = Coefficient of rolling friction between chain roller and load
μ_{S4} = Coefficient of sliding friction between chain roller and bush

Section	*Load pull assessment*
A—B	To move the chains on the unloaded strands

$$=wl_C\left(\frac{\mu_{R1}\times 2\times 12}{D}+\frac{\mu_{S4}d}{D}\right)$$

$$=wl_C\left(\frac{24\mu_{R1}+\mu_{S4}d}{D}\right) \tag{1}$$

B—C To turn the driven wheels

$$=0{\cdot}05wl_C\left(\frac{24\mu_{R1}+\mu_{S4}d}{D}\right) \tag{2}$$

C—D To move the chains on the loaded strands

$$=wl_C\left(\frac{24\mu_{R1}+\mu_{S4}d}{D}\right) \tag{3}$$

To move the dead load

$$=W\left(\frac{\mu_{R1}\times 2\times 12}{D}+\frac{\mu_{R3}\times 2\times 12}{D}\right)\times 2$$

$$=48W\left(\frac{\mu_{R1}+\mu_{R3}}{D}\right) \tag{4}$$

To overcome the impingement resistance of chain rollers against the bushes. This is the effort required resulting from the fact that the chain is literally spacing the load carrying rollers.

$$=48W\left(\frac{\mu_{R1}+\mu_{R3}}{D}\right)\times\frac{\mu_{S4}d}{D} \tag{5}$$

Total load pull

$$=(1)+(2)+(3)+(4)+(5) \text{ which simplified}$$

$$=48W\left(\frac{\mu_{R1}+\mu_{R3}}{D}\right)\left(1+\frac{\mu_{S4}d}{D}\right)+2{\cdot}05wl_C\left(\frac{24\mu_{R1}+\mu_{S4}d}{D}\right) \tag{6}$$

Rolling friction μ_{R1} for a steel roller on a rolled or pressed steel track is variable between 0·002 and 0·005 depending upon the track surface condition; it is normal to take the higher value. The friction between the roller and the load is also variable depending upon the latter. For the majority of applications it is sufficiently accurate as a rule to take μ_{R3} as also being equal to 0·005. Formula (6) will then become:

$$=48W\left(\frac{0{\cdot}005+0{\cdot}005}{D}\right)\left(1+\frac{\mu_{S4}d}{D}\right)+2{\cdot}05wl_C\left(\frac{24\times 0{\cdot}005+\mu_{S4}d}{D}\right)$$

$$=\frac{0{\cdot}48W}{D}\left(1+\frac{\mu_{S4}d}{D}\right)+2{\cdot}05wl_C\left(\frac{0{\cdot}12\times\mu_{S4}d}{D}\right) \quad \text{(lbf)}$$

Plate 50 Drum reclaiming plant

Two deep link plate chains are employed, widely spaced but connected at intervals by staybars fitted through link plate centres. The chain rollers run on support guides on the working strands but, on the return, outboard rollers carried on the projecting staybar ends serve to support the chains at intervals.

Plate 51 Steel mill transfer conveyor

A series of deep link plate chains form an ideal travelling bed in the handling of hot rolled steel joists. From final rolling to the straightening of rolls, it also acts as a cooling conveyor. Since the joists are conveyed purely by chain friction and pushers are not required, synchronisation between rows of chains is not essential and, in view of the considerable number of chains involved and the large overall width of the conveyor, this is a very important simplification.

Plate 52 Continuous steriliser for milk

This conveyor employs twelve strands of deep link conveyor chain to carry bottled milk in crates through a steriliser. Each strand of chain runs in a steel channel support and guide and the whole installation forms, in effect, a moving floor to the steriliser. The throughput is 1,000 gallons per hour, and the temperature is 107°C.

Plate 53 Core arbor cooling conveyor

The arbors, used in the casting of iron pipes, must be cooled quickly on removal from the moulds for re-use. The plate edge conveyor is, in this case, fitted with pusher dogs to allow the arbors to be conveyed up the inclined section. The chains carry the hot arbors through water with moulding sand in suspension, giving rise to corrosive and abrasive conditions with no possibility of lubrication. Accordingly, a high safety factor is used to yield a heavy chain with good bearing areas on all wearing components. Additionally, replaceable outboard rollers are provided to carry the load, leaving the main chain rollers to withstand only the wear due to wheel gearing action. Steel wheels are preferred in view of their resistance to abrasion and corrosion.

Plate 54 Crate conveyor

Crates are being conveyed by bearing their undersides directly on the chain rollers. They are constrained between guide strips on each side of the conveyor and each of the two chains run in a channel support and guide on the load carrying strand. The return strands could be similarly supported but in this case they are taken over flanged rollers, thus deriving a certain amount of automatic take-up by virtue of the catenary sag of the chains between pulleys.

Plate 55 Crate conveyor with cone bearing pins

Two conveyor chains running side by side form a crate conveyor in which the chains curve horizontally by virtue of coned bearing pins. Both load carrying and return strands of chain are supported on rolled steel angles, the vertical webs of which serve to guide the chains at the 5 ft radius bends. Continuous plates at each side of the conveyor guide the crates.

PASSENGER VEHICLE ASSEMBLY CONVEYOR

A large transport undertaking employs a special conveyor system to carry passenger vehicle chassis whilst re-assembling with the body after major overhaul. The conveyor comprises a continuous moving platform of concrete slabs 10 ft wide, 3 ft long and $2\frac{1}{2}$ in thick, carried on the rollers of four parallel chains, the inner pair of chains being driven to impart motion to the system. Side guides ensure that the slabs remain in line.

When each slab has traversed the 320 ft length of the system, it runs off the chains on to a roller bed and is then dropped 6 inches hydraulically and turned, in the horizontal plane, through a right angle. The slab is then pushed on to the return track which comprises two lines of rollers for supporting and two chains fitted at intervals with pusher dogs for propulsion purposes. These return the slabs at high speed, with their 'long' sides parallel to the main conveyor chains, underneath the conveyor bed to the starting end of the line where they are turned, raised and pushed back to the main conveyor by a second mechanism. The total load on the conveyor is of the order of 112 ton.

Plates 56 to 59 and schematic diagram illustrate the constructional details, layout and operation of the conveyor.

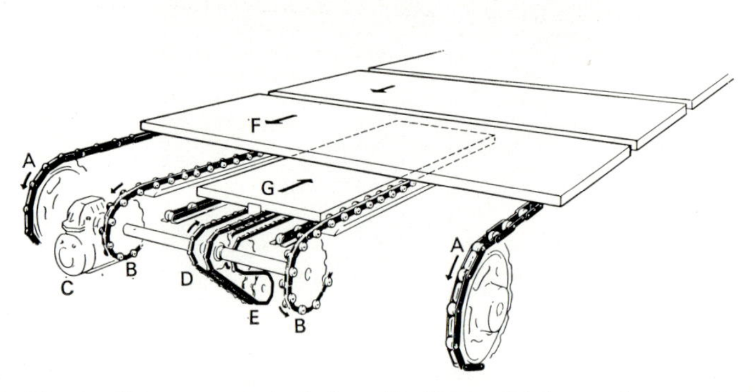

Plate 56 A. Outer support chains B. Inner driving and support chains C. Gearbox driving chains D. Return propelling chains E. Driving wheels for return propelling chains (gearbox not shown) F. Slab on conveyor G. Slab on return

Plate 57

View along the passenger vehicle conveyor with slabs removed, showing the four rows of slab support chains, the return roller tracks between the inner pair and the central return propulsion chains with dogs.

Plate 58 A slab across the conveyor

Plate 59 A slab in the return position

Plates 60 to 62 show a milk crate conveyor. This conveyor is installed to carry crates filled with bottles of milk through a mechanical crate stacker to the cool store to await distribution. A pair of deep link plate chains is employed to carry the crates on the plate edges and coned bearing pins are fitted to enable the chains to negotiate a right angle bend in the horizontal plane.

By use of the coned bearing pins the return strands may be brought in between the loaded strands as the diagram shows. As a result the return pit in the floor is of minimum depth with two resultant advantages, first that the cost of the steel trough in which the conveyor is built is minimised, and second that there is no interference with headroom in the basement located below the floor of the dairy.

As will be seen from Plate 62 the return strands are totally enclosed by a steel cover and they maintain their position at the point where the chain negotiates the bend.

Plate 60

Plate 61

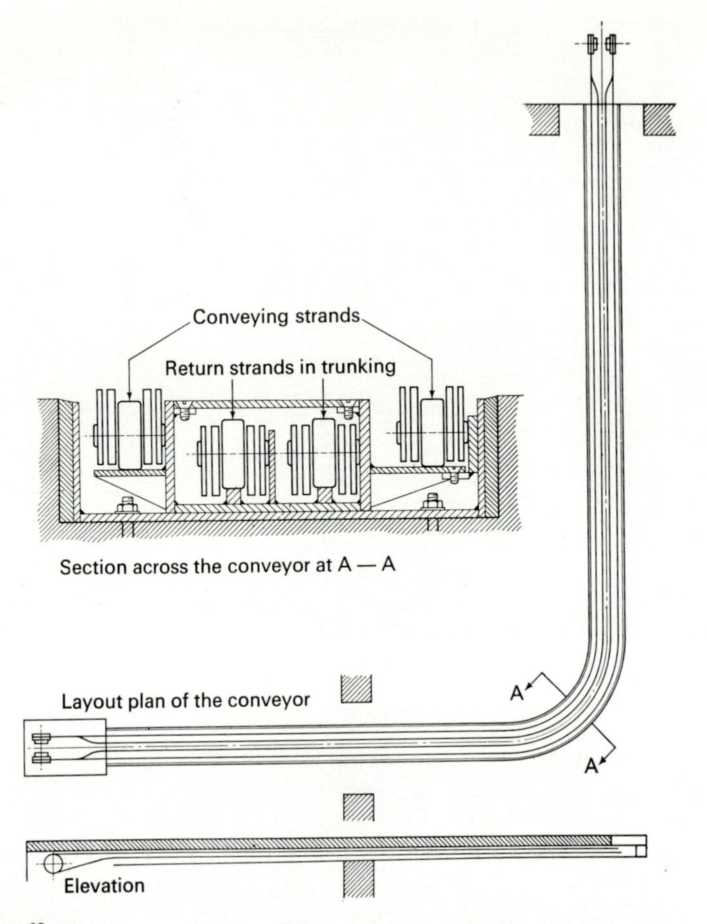

Plate 62

Reference to the cross sectional drawing shows how the loaded strands are carried by the chain rollers running on flat steel plating whereas, by reason of the projecting sideplates, the return strands call for narrow strip supports of less than the chain inside width. At the bend, the side load is taken by the bearing pin ends which slide against the steel rubbing faces provided at the sides of the floor channel and of the return strand casing. Alternatively the rubbing faces may be grooved so that the side load is taken by the chain plate and not the bearing pin.

10

Overhead Chain Conveyors

Description and chain types **10.1**

In its usual form the overhead chain conveyor embodies load supporting trolleys which are interconnected and moved by a single endless chain which acts purely as the traction medium. When operation is required in one plane, i.e. horizontal, the system is termed 'uniplanar'; when operating in horizontal and vertical planes the term 'biplanar' is applied.

It is one of the most versatile of conveyor systems since almost any desired circuit path may be followed. Circuit lengths are not limited by the strength of the chain used since it is possible to incorporate multiple drives which can be synchronised for correct load pull distribution within the respective sections. Although the main use is the transportation of goods it can also be used as a means of storage, a means of flow line co-ordination of machine output and as an integral part of particular processes such as degreasing, painting, drying etc. One of its major advantages is that it makes use of ceiling space which would be otherwise considered 'dead space'.

Many designs of overhead chain conveying systems are available and description of particular types follows.

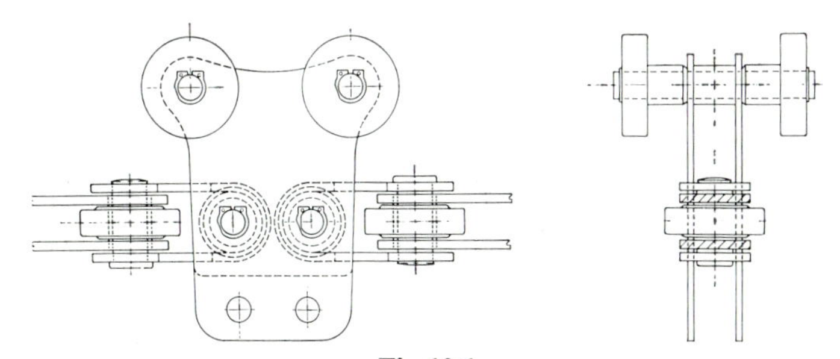

Fig 10.1

'VERTICHAIN', 4 ROLLER TYPE TROLLEY

With this type of chain, illustrated in Fig 10.1, a load can be carried vertically whereas trolleys with two rollers are normally limited to inclines not exceeding 45° from the horizontal. Normal trolley spacing is 24 in, the trolleys being connected by a 3 pitch length of chain which has biplanar stirrups at each end. The trolley is supplied loose as shown in Fig 10.2; this facilitates easy erection by enabling the trolleys to be positioned on the track and then coupled to the chain by connecting pins which are retained by soft circlips.

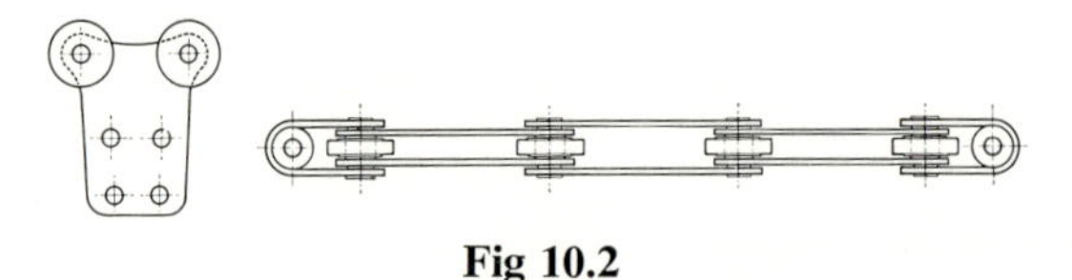

Fig 10.2

Two forms of track can be used with this chain; a fabricated track using rolled steel sections as shown in Figs 10.3 and 10.4 or a more sophisticated pressed steel track as Fig 10.5. With both types of track the chain and trolley rollers are guided and thus horizontal and vertical bends are negotiated without recourse to wheels. Minimum horizontal bend radius is 36 in. For vertical bends the minimum radius depends on trolley spacing, e.g. 24 in radius for trolleys at 12 in spacing and 36 in radius for 24 in trolley spacing. Typical bends are shown in Fig 10.6. On bottom vertical bends the track section is similar to that shown in Fig 10.3 but on top bends the depth of the vertical leg must be increased to provide a continuous guide for the chain rollers. Toothed wheels may be used at horizontal turns although normally this method is used only at drive positions. When using toothed wheels it is necessary to remove part of the track to allow free passage of the wheel as shown in Fig 10.7.

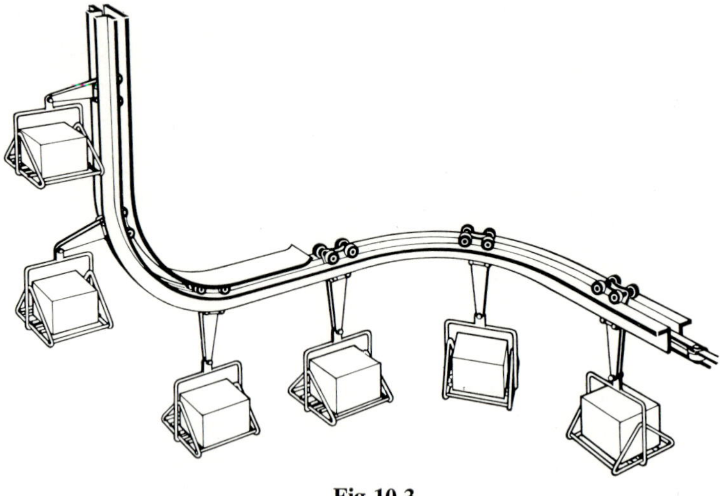

Fig 10.3

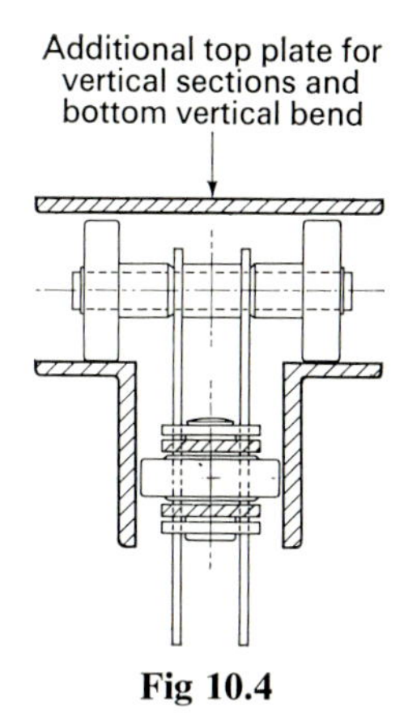

Fig 10.4

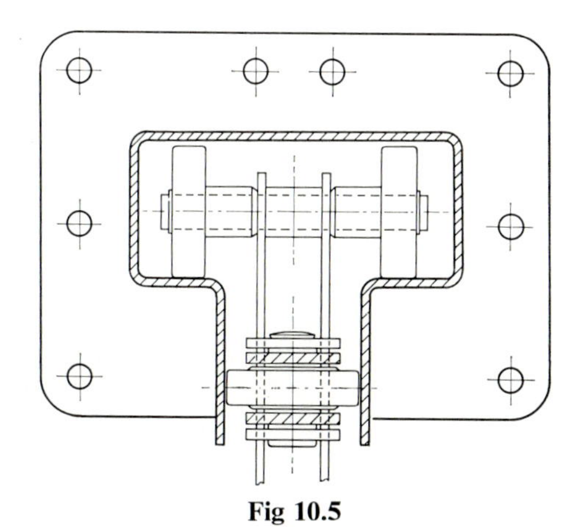

Fig 10.5

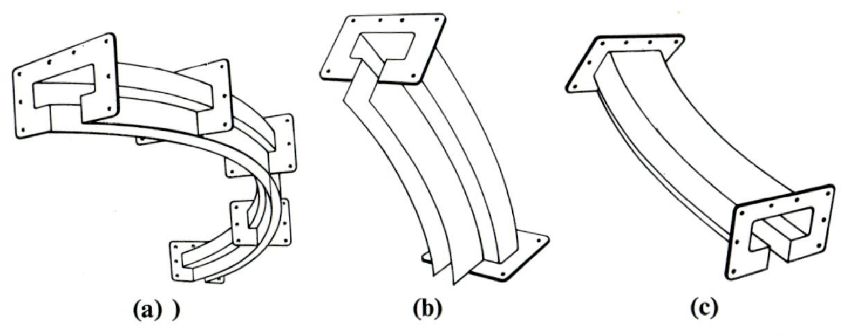

(a)) **(b)** **(c)**

Fig 10.6(a) Horizontal bend **(b)** Top vertical bend **(c)** Bottom vertical bend

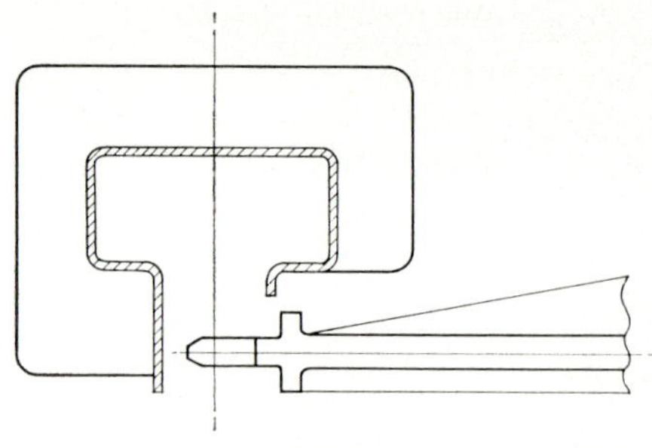

Fig 10.7

The chain as shown has a breaking strength of 15,000 lbf with a working load of 2,000 lbf. When selecting drive positions the effect of imposed loads on vertical bends must be considered. The imposed load plus suspended load on each trolley must not exceed 1,000 lbf. This is shown in Fig 10.8 and can be calculated as follows:—

$$\text{Imposed load (lbf)} = \frac{\text{Chain pull at bend (lbf)} \times \text{trolley spacing (in)}}{\text{Chain radius (in)}}$$

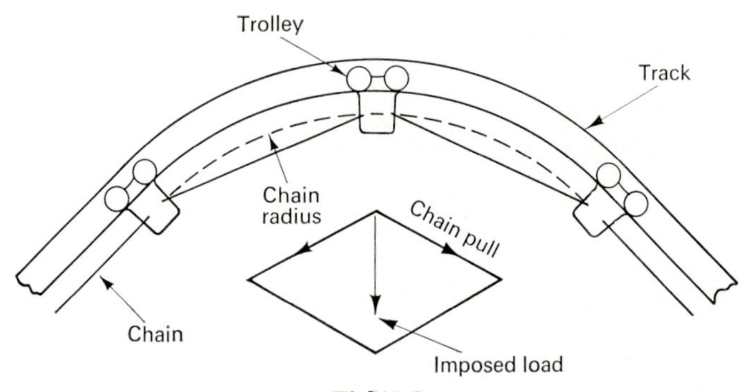

Fig 10.8

'TWO-TEN' CHAIN, 2 ROLLER TYPE TROLLEY

This overhead conveyor chain is shown in Fig 10.9. It is a precision chain which is an alternative to the drop forged rivetless chain in wide use, but with the following improved features:—

True pivotal movement in both planes.
All movement takes place between accurate hardened surfaces.
Full bearing contact at all angles of articulation.
Reduced frictional losses at pivot points.

Backlash is negligible.

Accurate chain pitch ensures that the most stringent pitching and timing requirements can be easily accommodated.

Chain assembly or disassembly is easy and positive.

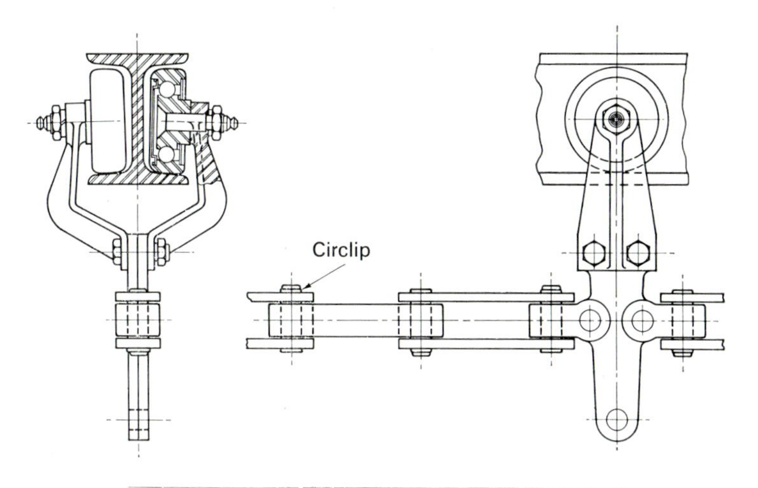

Vertical Bends—Min. Radius to Centre of Joist				
Trolley Spacing	16″	24″	32″	40″
Track Radius	24″	40″	56″	72″

Maximum chain pull = 3,000 lbf biplanar
Maximum chain pull = 3,500 lbf uniplanar
Trolley load = 400 lbf
Maximum resultant imposed load on bend = 800 lbf

Fig 10·9

STANDARD CHAIN, 2 ROLLER TYPE TROLLEY

6,000/7,500 lbf breaking load standard chain is attached to trolleys using biplanar joints as shown in Fig 10.10.

In each of these chains the trolleys are readily assembled to the chain thus facilitating easy site assembly. Trolley rollers incorporate ball bearings which are greased packed and fitted with nipples for subsequent greasing; provided that special attention is given to lubrication these rollers are suitable for temperatures up to 175°C. Roller peripheries are contoured to provide optimum contact with the track flanges. The inclusion of stirrup links enables small bend radii in the vertical plane to be negotiated.

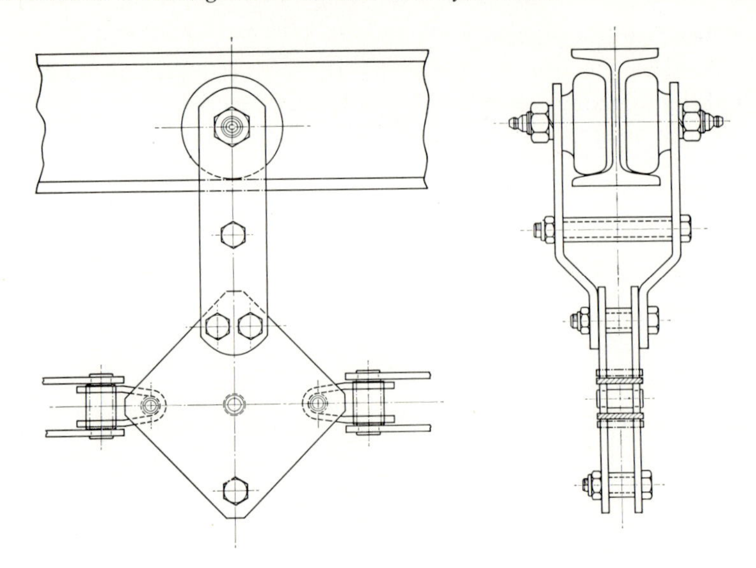

Vertical Bends—Min. Radius to Centre of Joist				
Trolley Spacing	12″	24″	36″	48″
Track Radius	24″	36″	72″	108″

Trolley load = 150 lbf
Maximum resultant imposed load on bend = 350 lbf

Fig 10.10

At horizontal bends the chain can be turned by three different methods.

Tooth wheel.

Plain wheel.

Roller turn.

Where the chain pull approaches the maximum working load it is preferable to use toothed wheels. For lap angles of 180° the wheels should have a minimum of 12 teeth although a greater number of teeth will promote smoother running. For lap angles less than 180° the aim should be to have at least 3 teeth in gearing engagement, for drive wheels the minimum number of teeth in engagement should be increased to 4.

The track above the wheels should have a radius equal to that of the inscribed circle of the polygon formed by the centre line of the chain links around the wheel, and should be blended with the straight tracks leading to and from the wheels. These straight tracks should be set midway between the radii of the wheel pitch circle and the polygon circle mentioned above.

Fig 10.11 illustrates this point but for wheels above 30 in diameter the difference between the track diameter and track centres becomes too small for practical consideration and therefore the track centre distance should be made the same as track diameter.

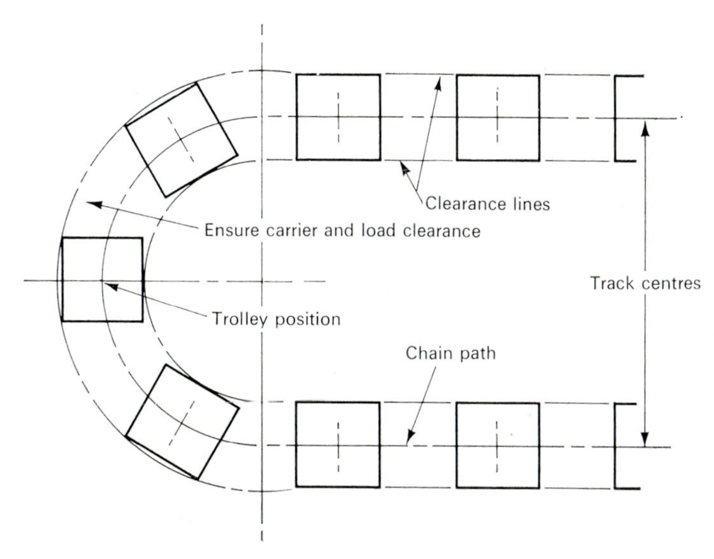

Fig 10.11

Plain wheels as shown in Plate 63 normally vary in diameter from 2 ft to 5 ft. These wheels cost less than toothed wheels but care must be taken when selecting the wheel diameter because of the bending effect on the chain plates. For example with a 22 in diameter wheel the pull should not exceed 2,500 lbf for the 'Two-Ten' chain. Where chain pull is high then the wheel should be as large as possible. Comments made for toothed wheels regarding positioning of tracking and wheel bearings also apply to plain wheels.

A roller turn as shown in Plate 64 can be used to turn the chain through any angle although it is generally employed where large bend radii are required or when the angle of lap is less than 90°.

Plate 63

Plate 64

Plate 65

Radius of the bend should be as large as possible especially when the chain pull approaches the maximum. Normally bend rollers have hardened treads and are fitted with anti-friction bearings. Spacing of the rollers should be as close as possible to provide smooth running of the chain.

For vertical bends the track must be curved as a true radius; the minimum radii for the various trolley spacings are shown in Figs 10.9 and 10.10. For this type of overhead chain the angle of inclination with the horizontal should not exceed 45° otherwise a toggling action may be induced into the chain since the line of the chain pull is offset from the point of load support (i.e. the trolley roller tread on the beam flange).

The use of toothed wheels with cut or cast teeth and plain wheels depends on the type of chain and the magnitude of the chain pull. Whichever type of wheel is used they should be fitted with anti-friction bearings to reduce the effect of bend friction which is cumulative throughout the circuit. Wheels should be rigidly mounted to cater for reaction load resulting from the chain pull. It is necessary on toothed wheels to remove teeth which otherwise would foul the trolley or solid inner links.

10.2 Conveyor layouts

The path of the conveyor must allow the loads being carried to clear walls and machinery etc. In addition clearance must be provided so that adjacent loads do not foul each other, this is illustrated in Figs 10.11 and 10.12.

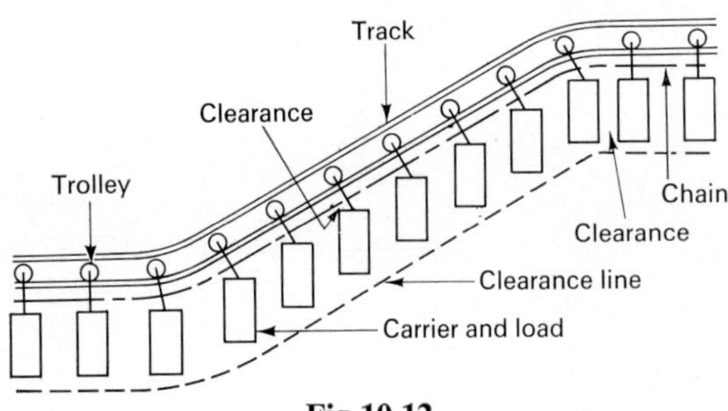

Fig 10.12

On uniplanar circuits the trolley spacing should not exceed 60 in and if load spacing in excess of this figure is required then intermediate trolleys will be necessary to support the chain so that this dimension is not exceeded. For biplanar circuits the trolley spacing should not exceed 42 in. Whichever system is used, uniplanar or biplanar, a trolley should be coincident with each load position to ensure that the load is borne directly by the trolley. Changes in elevation on an overhead circuit path are achieved by the use of curved tracks but when locating a vertical bend near a horizontal wheel or roller turn it is necessary to have a horizontal straight section of tracking equal to $1\frac{1}{2} \times$ trolley spacing between the wheel and the vertical bend to ensure that the chain remains horizontal whilst negotiating the bend.

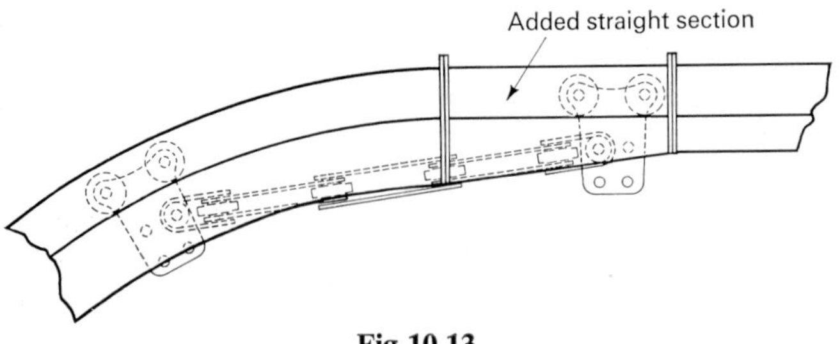

Fig 10.13

In the case of Vertichain where tracked bends are used it is necessary on top vertical bends to insert a straight length of tracking as shown in Fig 10.13 to ensure that the chain rollers are adequately guided. Horizontal and vertical bends should not be directly connected together, a 12 in length of straight tracking should be inserted between the bends.

Drive points should be located away from vertical bends particularly where the chain pull is high because of the reaction loading that is imposed.

Where the chain pull of an overhead conveyor is too great for the selected chain, additional drive points, a stronger chain or a circuit modification will be necessary. Additional drive units will also be required where the circuit contains a large number of bends even though the chain pull may not be high. As a general rule when a circuit contains more than 7–180° bend sections then additional drive points are suggested.

Provision must be made in the layout for chain adjustment otherwise excess slack will accumulate and chain bunching will occur. Apart from causing irregular motion this can also prejudice correct gearing action. The formula for allowable elongation (see Chapter 16) although not strictly correct will give satisfactory results.

Load carrying arrangements 10.3

When the load to be carried exceeds the allowable trolley loading it is common practice to join two trolleys together by a load bar, thereby sharing the load. Load bars may be used on biplanar or uniplanar systems.

With the arrangement as shown in Fig 10.14 the maximum rated capacity of each trolley may be employed. The load bar should be fixed so that variations in chordal to linear distance can be easily accommodated as bends or wheels are negotiated. For heavy loading two load bars are connected by a third bar as shown in Fig 10.15.

Tracking and supports 10.4

The size and type of tracking are as previously shown dependent on the chain being used. To assist in smooth running of the conveyor particular attention should be given to the alignment of track joints which should present a smooth surface to trolley rollers, otherwise there will be high peak loading and increased audibility. Track malalignment can cause fouling of the trolleys and rollers resulting in an increase in the natural tendency for the load to swing.

If joining of the track sections is done by welding, then to promote smooth running excess weld should be removed from the surfaces on which the rollers run or where clearance is required for passage of the trolleys. Strengthening members are frequently employed at joints but these must not cause any obstruction.

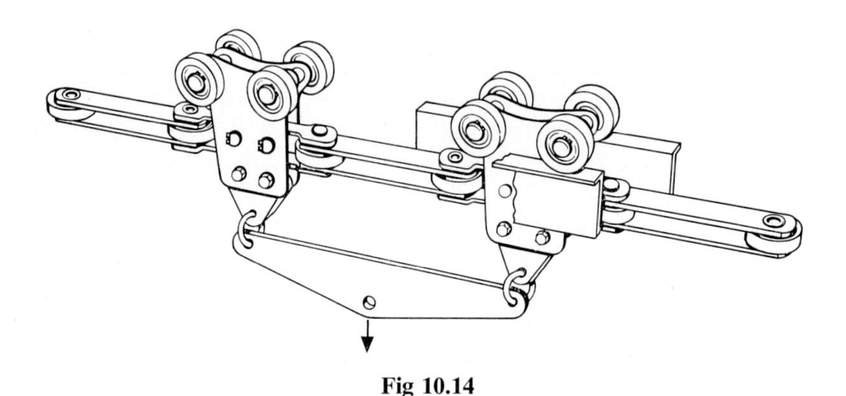

Fig 10.14

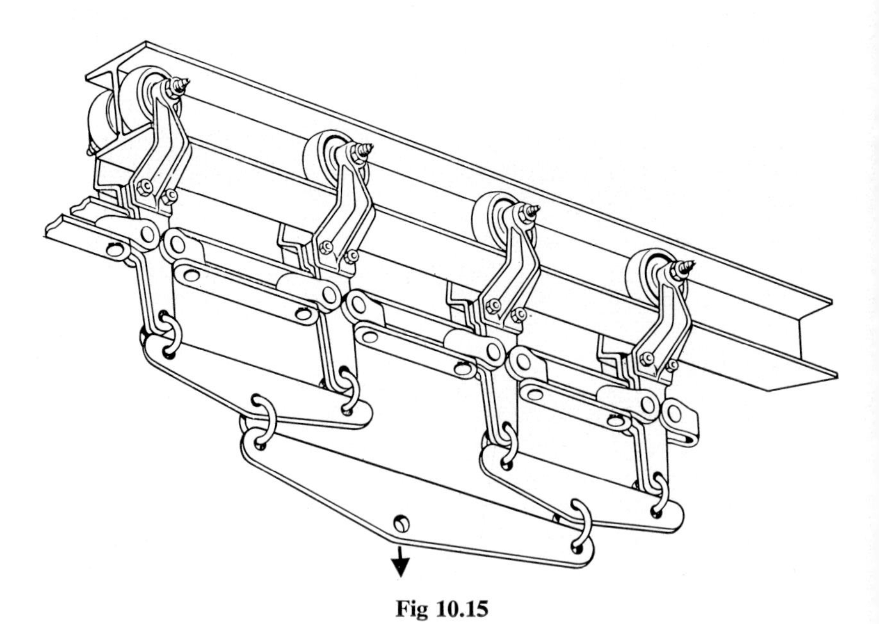

Fig 10.15

The track at vertical bends is subjected to two loads, the trolley load and the imposed load and although trolleys with ball bearing trolley rollers have high load capacities these should not be used to determine the allowable trolley load. The main factor limiting the load is what the track can withstand without excessive wear or peening and the rating given for the trolleys illustrated take this factor into consideration, provided that the trolleys operate on the recommended tracking.

At bend sections there will always be an inward reaction loading resulting from the load pulls in each section of the chain about the bend and therefore all structural work, bearings etc., should be rigidly mounted to contend with this loading so that distortion or deflection is negligible.

Track supports should be arranged so that no undue deflection of the track occurs under maximum live load conditions. Supports are normally taken from roof trusses, walls or building stanchions. On occasions floor stands are used but these should allow free passage of the chain, trolley and load.

As discussed earlier it is essential to cater for chain extension and therefore it is necessary to provide a means of adjusting the track length. These tension units, preferably telescopic, should provide a smooth and continuous support for the chain and trolley rollers. Normally tension units are arranged to operate in pairs and coincide with 180° chain lap.

Chain pull calculations 10.5

When calculating the chain pull for a circuit by means of the section method, it must be borne in mind that at horizontal bends two components of load pull exist as below.

(a) The effort required to rotate the wheel against bearing friction. As a general rule this can be taken as 3% of the chain pull at exit of the bend for 90° turns, and 5% for 180° turns. In the case of tracked bends where the chain rollers run on the tracking then the factor for various lap angles is shown in Table 10.1.

(b) The effort required to pull the chain and load around the bend. Fig 10.16 illustrates the method of calculating the pull at horizontal bends.

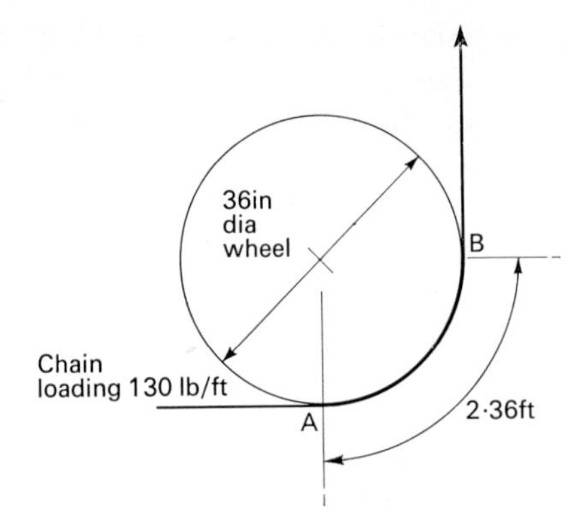

Pull at A = 1,000 lbf
Section AB = 2·36 × 130 × μ_{R4} (where μ_{R4} = 0·035)
= 10·7 lbf
Therefore pull at B = 1,010·7 × 1·03
= 1,041 lbf

Fig 10.16

As will be seen condition (a) has a cumulative effect around the circuit and therefore it is essential to determine the direction of chain travel before starting chain pull calculations.

In biplanar circuits the chain pull comprises (a), (b) and in addition (c) to negotiate inclines.

In the case of (c) a downward inclined section will only induce a positive component load if the trolley roller friction expressed in degrees, exceeds the track inclination θ to the horizontal (i.e. when $\mu_{R4}\text{Tan}^{-1}$ is greater than θ) (see Chapter 2).

On straight inclined sections the effort required to move the load depends on the degree of incline as shown in Table 10.2.

The calculation for vertical curves is similar to that previously described for horizontal bends. In this case the trolley rollers run on the track, and factors for the various angles of bends are shown in Table 10.3. Construction of vertical bend angle is shown in Fig 10.17.

In addition to the effort required to negotiate the bend there is also the effort required to raise the load. Fig 10.17 illustrates the method of calculation.

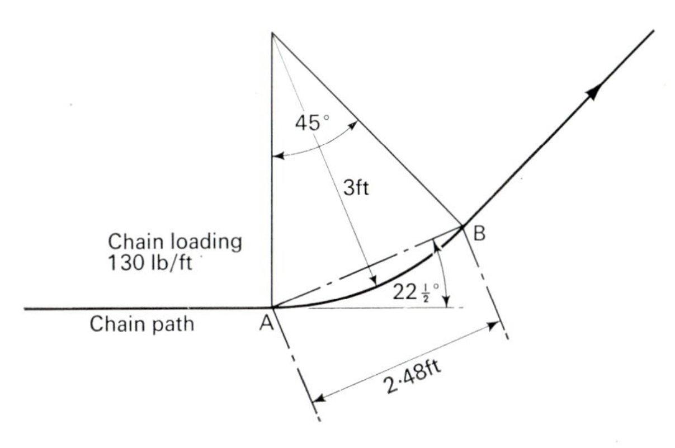

Pull at A = 1,000 lbf
Section AB = 2·48 × 130 × 0·42
= 135 lbf
Therefore pull at B = 1,135 × 1·028
= 1,165 lbf

Fig 10.17

An approximate selection method is available as shown below, but biplanar circuits normally require solving with the section method.

Chain pull (lbf) $= 1{\cdot}2\mu_{R4}w_C l_A + w_C h_C$ (Ignore $w_C h_C$ if rise and fall balance)

Where $1{\cdot}2$ = Approximate factor for circuit bends
μ_{R4} = Coefficient of friction of trolley rollers
w_C = Weight of chain, attachments and load (lb/ft)
l_A = Length of chain in circuit (ft)
h_C = Height from low to high level (ft)

EXAMPLE

An overhead conveyor 180 ft long as shown in Fig 10.18 is to handle castings each weighing 260 lb, spaced 16 in apart. The chain speed is to be 30 ft per min. Load and unload points are coincident.

For this application the use of the 'Two-Ten' chain is envisaged with a caterpillar drive.

It is required to determine:—

Position and number of drives
Chain pull
Maximum imposed load on vertical bends and bend radius
Running horsepower

In the tortuous circuit shown it is difficult to place the drive away from a bend and therefore other factors will have to be considered.

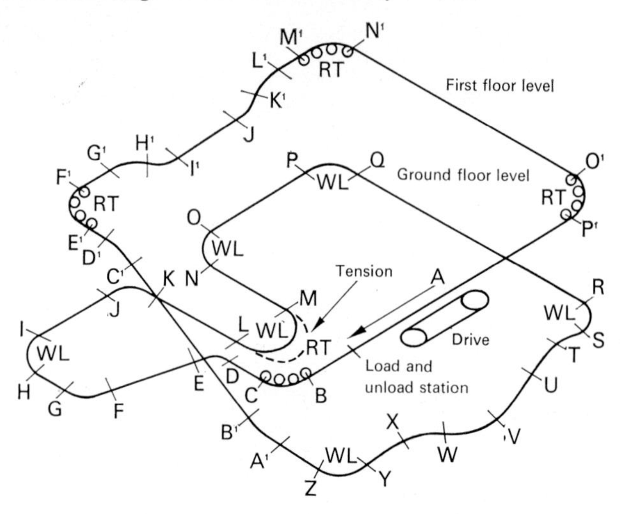

Fig 10.18 Isometric circuit line diagram (fully loaded circuit). WL denotes Plain Wheel turn, RT denotes Roller turn

It is preferable to have the drive at the 'high level' of the circuit, so that the chain pulls the load up the inclines; the drive should be located near the off-loading position. In the example shown the optimum position for the drive is at point A.

A very approximate calculation reveals that the proposed chain with one drive will be suitable, as shown below:—

Weight of chain and trolleys	$=8{\cdot}3$ lb/ft
Load $\frac{260}{16}\times 12$	$=195$ lb/ft
Loaded chain	$=203{\cdot}3$ lb/ft
Approximate load pull	$=203{\cdot}3\times 0{\cdot}035\times 180\times 1{\cdot}2$
	$=1{,}540$ lbf

It should be noted that as the rising and falling sections are equally loaded then these can be assumed to balance each other.

The final calculations can now proceed as follows.

DRAG FACTORS

Horizontal, fully loaded	$203{\cdot}3\times 0{\cdot}035$	=	7·1 lb/ft
15° rise, fully loaded	$203{\cdot}3\times 0{\cdot}293$	=	59·6 lb/ft
15° fall, fully loaded	$-203{\cdot}3\times 0{\cdot}224$	=	− 45·5 lb/ft
$22\frac{1}{2}$° rise, fully loaded	$203{\cdot}3\times 0{\cdot}42$	=	85·4 lb/ft
$22\frac{1}{2}$° fall, fully loaded	$-203{\cdot}3\times 0{\cdot}35$	=	− 71·2 lb/ft
30° rise, fully loaded	$203{\cdot}3\times 0{\cdot}53$	=	107·7 lb/ft
30° fall, fully loaded	$-203{\cdot}3\times 0{\cdot}47$	=	− 95·5 lb/ft

Section	Length (ft)	Bend	Factor	Pull (lbf)	Total pull (lbf)
A B	8·75		7·1	62·1	62·1
B C	3·14	90° roller turn	7·1 1·03*	22·3	84·4 86·9
C D	1		7·1	7·1	94·0
D E	2·07 at 15° fall	30° Vert (down)	−45·5 1·019*	−94·2	zero pull −0·2 (0) Est. pull to turnwheel =1
E F	10·4 at 30° fall		−95·5	−993·2	zero pull −992·2(0)
F G	2·07 at 15° fall	30° Vert (down)	−45·5 1·019*	−94·2	zero pull −94·2(0) Est. pull to turnwheel = 1
G H	1		7·1	7·1	8·1
H I	3·14	90° wheel turn	7·1 1·03*	22·3	30·4 31·3
I J	4·25		7·1	30·2	61·5
J K	3·14	90° wheel turn	7·1 1·03*	22·3	83·8 86·3
K L	5·5		7·1	39·1	125·4
L M	6·28	180° wheel turn	7·1 1·05*	44·6	170·0 178·5
M N	5·5		7·1	39·1	217·6
N O	3·14	90° wheel turn	7·1 1·03*	22·3	239·9 247·1
O P	5·5		7·1	39·1	286·2
P Q	3·14	90° wheel turn	7·1 1·03*	22·3	308·5 317·8
Q R	15		7·1	106·5	424·3
R S	3·14	90° wheel turn	7·1 1·03*	22·3	446·6 460
S T	1		7·1	7·1	467·1
T U	3·06 at 22½° fall	45° Vert (down)	−71·2 1·028*	−217·8	249·3 256·3

* Bend drag factor

Section		Length (ft)	Bend	Factor	Pull (lbf)	Total pull (lbf)
U	V	3·06 at 22½° fall	45° Vert (down)	−71·2 1·028*	−217·8	38·5 39·6
V	W	3·06 at 22½° rise	45° Vert (up)	85·4 1·028*	261·3	300·9 309·3
W	X	3·06 at 22½° rise	45° Vert (up)	85·4 1·028*	261·3	570·6 586·6
X	Y	1		7·1	7·1	593·7
Y	Z	3·14	90° wheel turn	7·1 1·03*	22·3	616·0 634·5
Z	A¹	1		7·1	7·1	641·6
A¹	B¹	2·07 at 15° rise	30° Vert (up)	59·6 1·019*	123·4	765·0 779·5
B¹	C¹	10·4 at 30° rise		107·7	1,117·1	1,896·6
C¹	D¹	2·07 at 15° rise	30° Vert (up)	59·6 1·019*	123·4	2,020·0 2,058·4
D¹	E¹	1		7·1	7·1	2,065·5
E¹	F¹	3·14	90° Roller turn	7·1 1·03*	22·3	2,087·8 2,150·4
F¹	G¹	1		7·1	7·1	2,157·5
G¹	H¹	3·06 at 22½° fall	45° Vert (down)	−71·2 1·028*	−217·8	1,939·7 1,994
H¹	I¹	3·06 at 22½° fall	45° Vert (down)	−71·2 1·028*	−217·8	1,776·2 1,826
I¹	J¹	2·25		7·1	16	1,842
J¹	K¹	3·06 at 22½° rise	45° Vert (up)	85·4 1·028*	261·3	2,103·3 2,162
K¹	L¹	3·06 at 22½° rise	45° Vert (up)	85·4 1·028*	261·3	2,423·3 2,491
L¹	M¹	1		7·1	7·1	2,498·1
M¹	N¹	3·14	90° Roller turn	7·1 1·03*	22·3	2,520·4 2,596
N¹	O¹	15		7·1	106·5	2,702·5
O¹	P¹	3·14	90° Roller turn	7·1 1·03*	22·3	2,724·8 2,807

* Bend drag factor

Section		Length (ft)	Bend	Factor	Pull (lb)	Total pull (lbf)
P^1	A	8·75		7·1	62	2,869

$$\text{Maximum chain pull} = 2{,}869 - (0{\cdot}2 + 992{\cdot}2 + 94{\cdot}2)\ \text{lbf}$$

$$\text{Running hp} = \frac{[2{,}869 - (0{\cdot}2 + 992{\cdot}2 + 94{\cdot}2)] \times 30}{33{,}000}$$

$$= 1{\cdot}61$$

The allowable resultant imposed load for the trolley is 800 lbf as shown in Fig 10.9 and therefore the minimum radius for the vertical bend at K^1L^1 can be calculated as follows

$$800 = \frac{2{,}491 \times 16}{\text{Chain radius}} + 260$$

$$\text{Therefore chain radius} = \frac{2{,}491 \times 16}{800 - 260} = 74\ \text{in}$$

Overhead conveyor drives 10.6

On overhead conveyors chain speed is normally low which tends to promote frictional conditions that may cause uneven running of the chain. On vertical bends the chain forms a series of chords and variations in the chain length can produce speed and load variation as the chain slackens and tightens.

Toothed wheels are widely used for driving, and wherever possible these should accommodate a 180° chain lap. With 12 teeth, satisfactory smoothness of running will be obtained, but obviously any increased numbers of teeth is advantageous. When the angle of lap is below 180° the aim should be to have a minimum of 4 teeth in engagement with the chain. The wheels operate horizontally (i.e. with shafts vertical) and must be gapped to clear the trolleys, the number of teeth being a multiple of the trolley spacing in pitches.

On overhead conveyors with a complex track formation it is quite possible for two sections of the circuit to be in close proximity. In such a case, two or even more power take-offs can be arranged from one power source. Plate 66 shows an example of this, using a single electric motor. It should be noted that the strand of conveyor chain on the left deviates from the straight to incorporate a 180° drive point. On the right-hand side, the originally planned U-bend was altered to a V formation to include a second driver. Two identical chain drives from the motor complete the arrangement, the driving chains being individually adjusted.

Caterpillar drives are used where a toothed drive wheel cannot be accommodated. Fig 10.19 shows a caterpillar drive designed specifically for use with Vertichain.

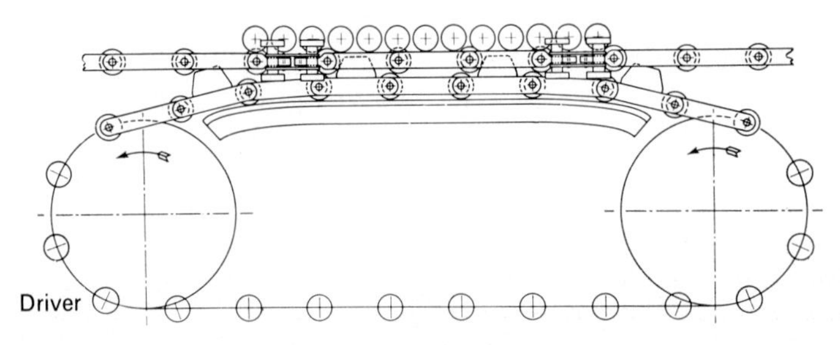

Fig 10.19

Plate 66

10.7 Overhead towing systems

Overhead systems may be used for towing trucks etc., over a predetermined path via a connecting member from the chain trolley to the truck. This member may be non rigid (Fig 10.20) such as rope or chain, alternatively a rigid member permanently fixed to the truck and manually engaged or disengaged with the chain trolley. When the trucks run on one level then the non rigid connection may be used. The trucks may be guided by running the wheels on tracks or allowed to freely follow the towed path. When a rigid connection is used both upward and downward inclines may be traversed, the connector being capable of 'hold back' in both directions.

Wheeled trucks exhibit peculiar movement characteristics when traversing horizontal bends especially if all or certain of the wheels are castored. A good deal of space can be 'lost' in allowing sufficient space for unimpeded truck turning. To avoid this waste of space it is necessary to introduce a compounded curve for the chain path at a bend section to guide the truck through a minimum sweep area. Such is the case illustrated in Plate 67.

Plate 67

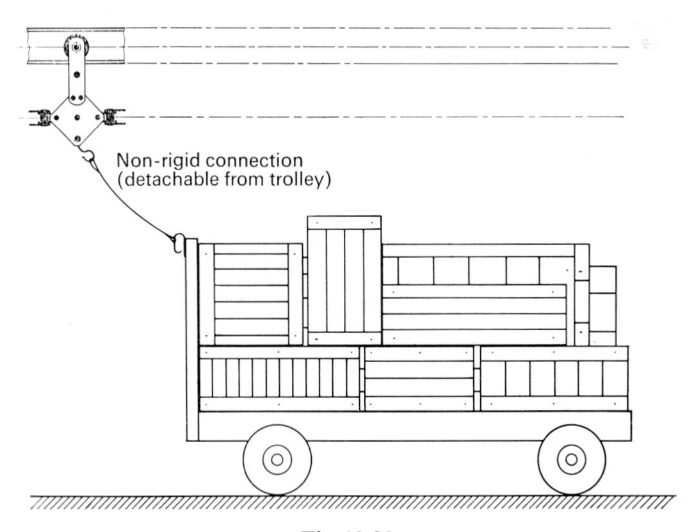

Fig 10.20

Here at a right angled turn, the chain is constrained to follow a double curve, Fig 10.21, with the four castored truck wheels following the floor path as illustrated.

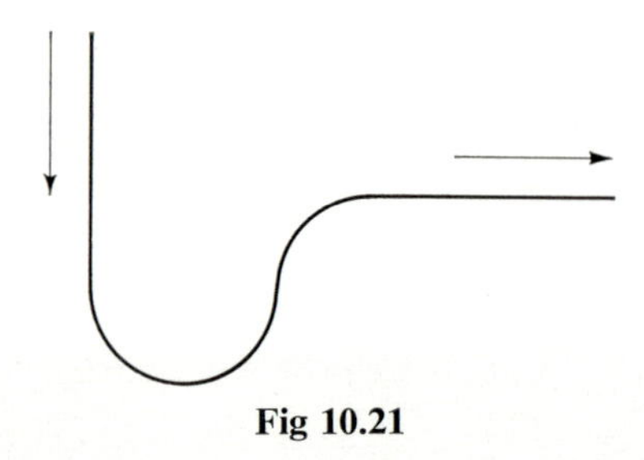

Fig 10.21

The compounding of the curve is dependent upon the individual truck characteristics, such as wheel location, pitching, towing point relative to the wheels and to the centre of gravity of the loaded truck.

Towed truck systems with rigid connectors can be engineered to both accept and release the trucks automatically at predetermined points. It is usual to have safety devices which will quickly release the trucks in the event of contact with an obstruction or personnel.

The alternative truck movement system is the underfloor tow conveyor as previously described under Chapter 8.

10.8 Power and free conveyors

Pusher conveyors using a driven overhead system to propel independently supported load carrying trolleys are extensively used. The system affords the facility of switching or transferring onto an auxiliary line, at any desired position for work operations or storage. Tracks for the load carrying trolleys mounted below the powered chain track are connected by a common yoke. To effect trolley release the powered chain is caused to rise and run above the trolley track, thereby necessitating the use of a biplanar chain. The trolley may also be released at a horizontal turn in the lower track. Load trolley designs vary considerably depending upon the duty involved. Influencing factors include load to be carried, circuit bend and switch line radii, and the overall requirements of engagement and disengagement with the powered chain attachment. On this latter aspect the propelling dog can either be incorporated in the powered chain, with a suitable mating projection on the load trolley, or the dogs are mounted on the trolleys contacting fixed projections on the powered chain. For the negotiation of inclines of some magnitude, whilst still under control of the propelling chain, trolleys with 4 wheels are normally used for actual load carrying with additional wheels or rollers for lateral constraint. A simple adaptation of standard overhead chain is indicated in Fig 10.22 where the hinged pusher allows in-feed of the load in the direction of chain movement.

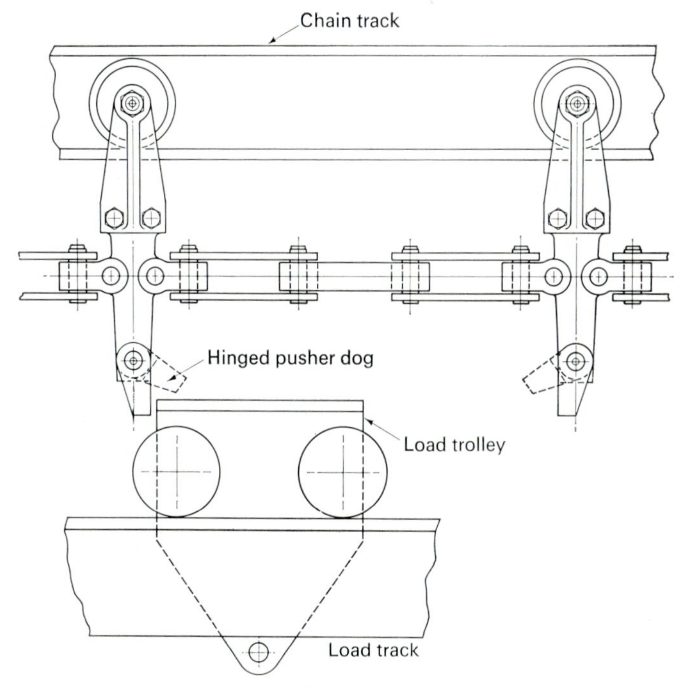

Fig 10.22

This can be applied equally well to the use of the stirrup type biplanar chain and also to Vertichain (Fig 10.23).

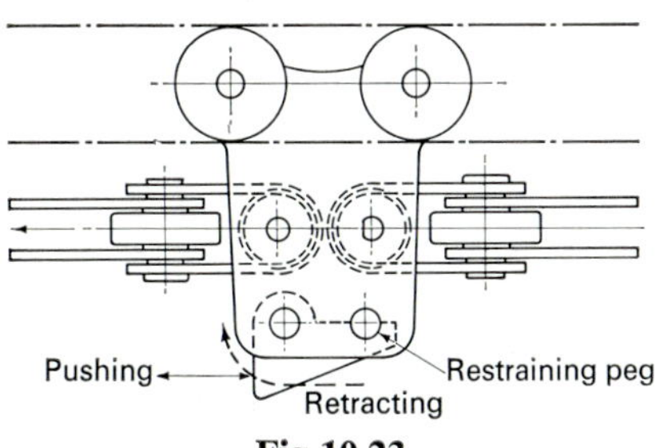

Fig 10.23

Standard conveyor chain with simple attachments may also be adapted for overhead work, as is illustrated in Plates 68 and 69 where a hinged pusher provides a retracting feature.

Plate 68

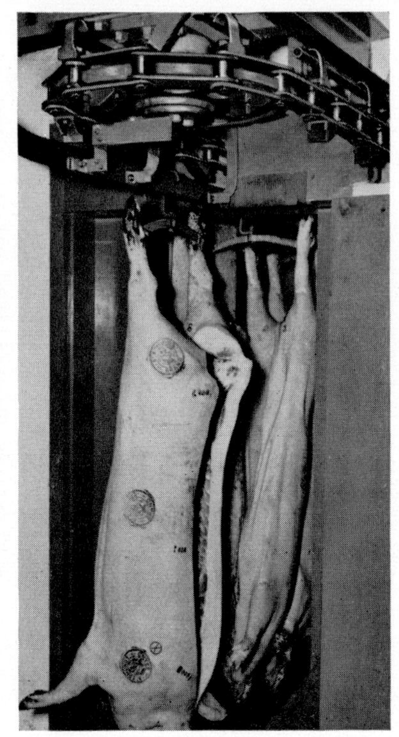

Plate 69

Plates 68 and 69 Carcase cooling conveyor in a bacon factory

Carcases are hooked onto a 'load rail' and propelled by pusher attachments fitted to the chain, through a chilling room to the pickling operation. 7·0 in pitch solid bearing pin bush chain of 45,000 lbf breaking load is employed with attachments every outer link. The loaded conveyor has a capacity of 300 carcases.

A further arrangement is shown in Fig 10.24, in which case chain support is from a plain plate sliding on angle tracking, side guiding is obtained via the chain rollers. The loads would be hung so that their centre of gravity coincides with the centre of the track.

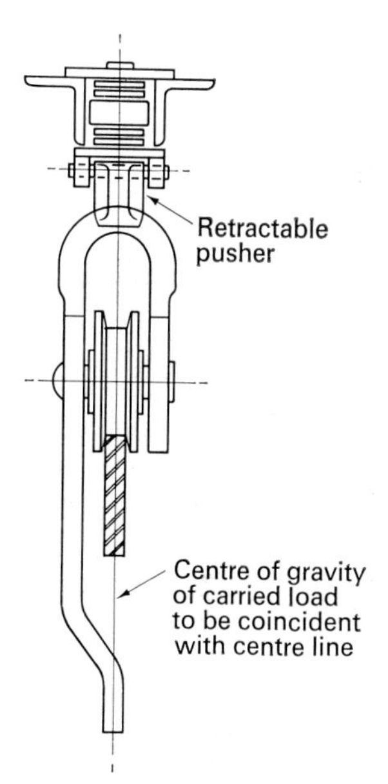
Retractable
pusher
Centre of gravity
of carried load
to be coincident
with centre line

Fig 10.24

11

Vertical Elevators

The elevating systems detailed in this chapter comprise single or multiple chain layouts, in which the chain functions mainly as a lifting medium, but may also be required to perform horizontal conveying within the one circuit.

11.1 Swing tray elevators

Fig 11.1

This type of elevator shown in Fig 11.1, and Plates 70, 71 and 72 is used for carrying unit or packaged loads, in trays which pivot between two parallel chains. The method of attachment allows the tray carrying surface to remain horizontal at all times, regardless of the chain path.

The trays should be rigidly constructed to ensure that the chains are tied together so as to remain parallel and obviate canting. Automatic loading and unloading can be arranged as shown in Figs 11.2 to 11.4 by the use of grid platforms, the platform base having a shaped cutaway through which the trays can pass.

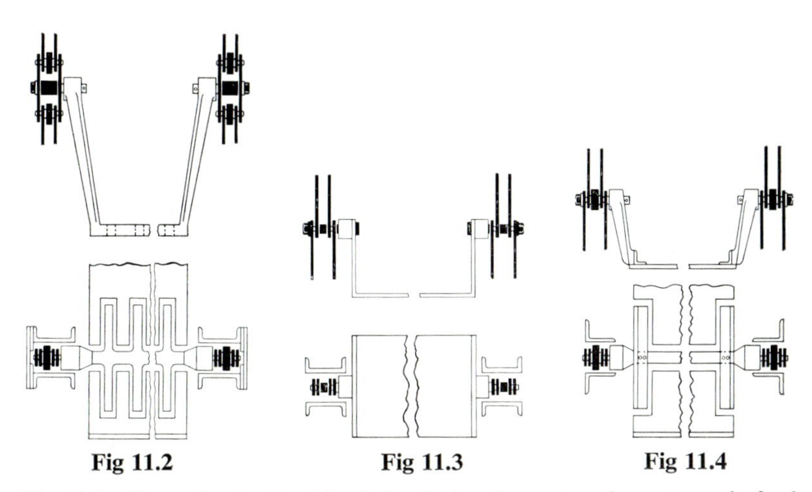

Fig 11.2 **Fig 11.3** **Fig 11.4**

Fig 11.2 Heavy loads. Double chain. Swing finger tray for automatic feed and discharge by means of grid loading and unloading platforms.

Fig 11.3 Light loads. Double chain. Swing tray for hand feed and discharge.

Fig 11.4 Medium and light loads. Double chain. Swing finger tray for automatic feed and discharge.

If a single headshaft is employed the wheel must allow free passage of trays without fouling the shaft and wheel bosses. When handling large loads it may be necessary for stub shafts to be used, in which case the headwheels must be driven from a common countershaft and correctly synchronised, in this connection reference should be made to Chapter 15.

Any tendency of the chains or trays to swing must be resisted, and on most installations guide tracks are employed. Depending on the combined weight of chains, trays and load, it may be necessary for semi-circular guides to be fitted close to the tail wheel periphery otherwise the chain may drop away from the wheel.

The chain pitch is generally dictated by the required tray spacing, along with the correct selection of number of wheel teeth. The pivotal attachment of the tray to the chain can consist of an extended bearing pin or spigot pin which is rigidly fixed through the chain plates or hollow bearing pins. Whatever form is used, the fitting must allow free pivoting of the tray as shown in Figs 11.5 and 11.6. As load pick-up may introduce shock loading, the pivot pin strength must be sufficient to contend with this.

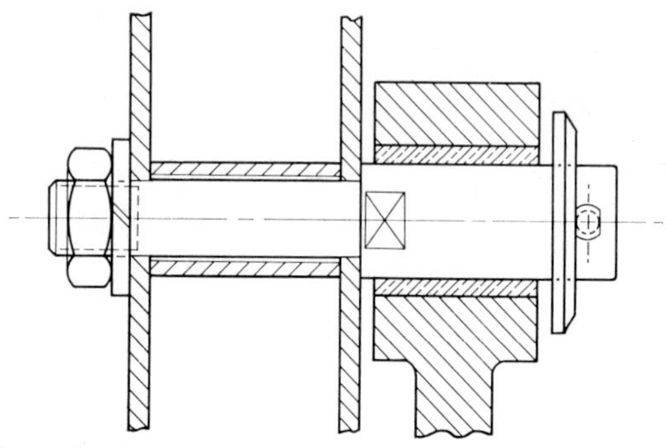

Fig 11.5 Spigot pin through outer link provides tray pivot point

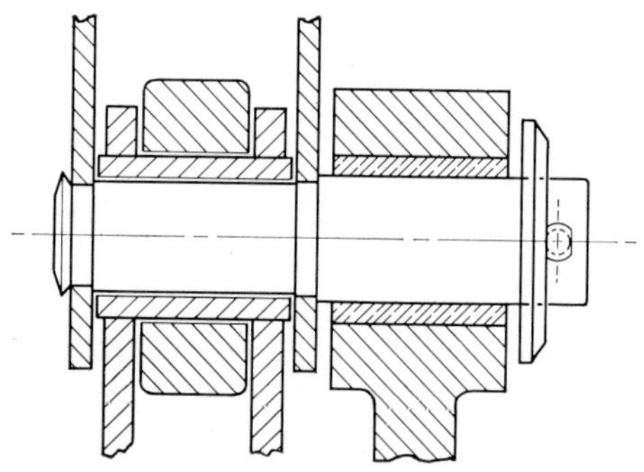

Fig 11.6 Tray pivot point using an extended bearing pin

For most applications wheels having a cast tooth form are suitable, the number of teeth being the usual standard minimum of 8. Gapped teeth are necessary if the trays are supported by spigot pins passing through the chain plates.

On simple layouts as Fig 11.1 adjustment is normally achieved at the tailshaft end, provision must be made for securely locking the adjustment after re-positioning. In cases where adjustment at the tailshaft cannot be arranged, this must be carried out at the headshaft. For complex layouts the wheel directly following the driver, i.e. at the lowest point of chain pull, is employed for take-up purposes. This is shown in Fig 11.7, where the best position would be at DE, maximum adjustment being obtained with minimum wheel movement. At the same time, the vertical relationship between chain strands CD and EF would be retained.

If a number of drives are used for one circuit it is normally preferable to provide a separate adjustment portion for each drive. Maintenance of correct adjustment is essential, since otherwise excessive chain slack is likely to allow the trays to swing with consequent spillage.

Chain pull calculations — 11.2

The method of calculation is best illustrated by working through an example.

Referring to Fig 11.7:—

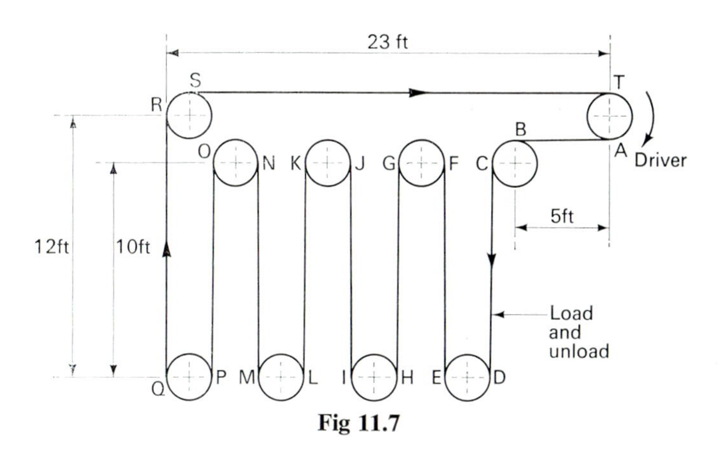

Fig 11.7

Equal loading conditions are obtained throughout, and thus the vertical chain strands (which are of identical height) will be in equilibrium about the wheel positions FG, JK and NO. There will also be a tendency at all lower wheel positions, for the chain to drop away from the wheels.

Where w_C = Weight of material, chain and attachments = 30 lb/ft

μ_R = Overall coefficient of friction of chain rolling on track = 0·15

S = Chain speed = 60 ft/min

Horizontal drag factor = 30 × 0·15 = 4·5 lb/ft

Vertical drag factor = 30 lb/ft

Calculations are commenced at position A where the load pull is theoretically zero. The pull at C will be the pull over section AB plus a 3% increase to allow for 90° negotiation of the wheel. In section CD the chain will fall due to its own weight, thereby imparting an assisting pull to the drive wheel without increasing the maximum pull which occurs at T. As the pull in section CD is greater than that at C the resultant pull will be a negative quantity and therefore for practical purposes calculations are recommenced at D, the assisting pull being taken into consideration when the horsepower requirement is calculated.

On recommencement at D it is necessary to estimate the pull required for wheel negotiation. The pull at F will be the pull over section EF added to the pull at E, and at G the pull will be increased by 5% to allow for 180° negotiation of the wheel. Due to the fall in the circuit between G and H the chain pull at H is reduced.

A similar approach is adopted throughout the circuit until point T is reached, where, as shown, the maximum chain pull is 529·6 lbf.

Section	Length (ft)	Wheel	Factor	Pull (lbf)	Total pull (lbf)
AB	5		4·5	22·5	22·5
BC		90°	1·03		23·2
CD	10		−30	−300	−276·8(0)
DE		180°			Estimate 1
EF	10		30	300	301
FG		180°	1·05		316·1
GH	10		−30	−300	16·1
HI		180°	1·05		16·9
IJ	10		30	300	316·9
JK		180°	1·05		332·7
KL	10		−30	−300	32·7
LM		180°	1·05		34·4
MN	10		30	300	334·4

Section	Length (ft)	Wheel	Factor	Pull (lbf)	Total pull (lbf)
NO		180°	1·05		351·1
OP	10		−30	−300	51·1
PQ		180°	1·05		53·7
QR	12		30	360	413·7
RS		90°	1·03		426·1
ST	23		4·5	103·5	529·6

$$\text{Running horsepower} = \frac{(529{\cdot}6 - 276{\cdot}8) \times 60}{33{,}000}$$
$$= 0{\cdot}46$$

Finger tray elevator **11.2**

This is a form of swing tray elevator which uses only a single chain strand as shown in Figs 11.8 and 11.9; representative installations are shown in Plates 73 and 74.

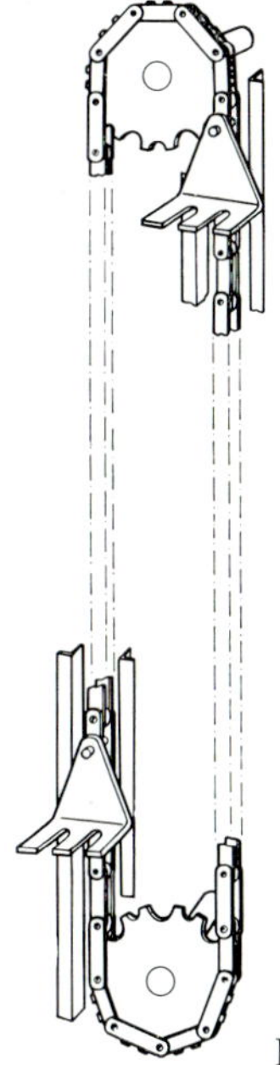

Fig 11.8

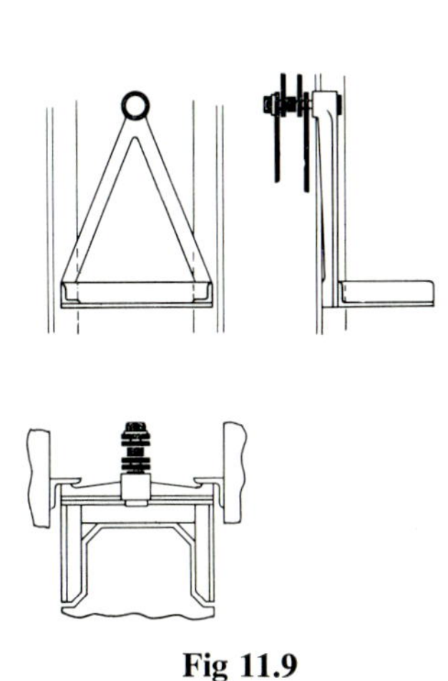

Fig 11.9

The offset loading introduced by the method of supporting the carriers from the chain is counteracted by allowing the back of the carrier to bear against a continuous smooth guide track. In some cases anti friction rollers are used to reduce friction and wear. Virtually any desired circuit path may be accommodated.

CHAIN PULL CALCULATIONS

This is calculated in accordance with Chapter 2 but allowance is necessary for the additional load pull induced by the offset load and tray. Referring to Fig 11.10:—

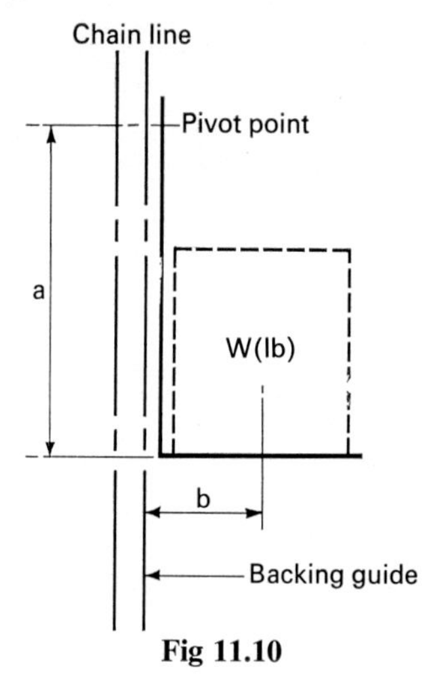

Fig 11.10

Where W_D = Weight of material on tray (lb)

W_C = Weight of tray (lb)

a = Distance from pivot point to contact position of tray and backing guide (in)

b_L = Distance of centre of gravity of carried load from backing guide (in)

b_T = Distance of centre of gravity of tray from backing guide (in)

μ_{S5} = Coefficient of friction between tray and backing guide (may be sliding or rolling depending upon design)

Additional load pull to be taken into account per tray position

$$= \mu_{S5}\left(\frac{W_D b_L}{a} + \frac{W_C b_T}{a}\right)$$

$$= \mu_{S5}\left(\frac{W_D b_L + W_C b_T}{a}\right)$$

The horsepower assessment is made in line with Chapter 2 but with allowance for the additional load pull as in the foregoing, and noting that it is the unbalanced pull which is of significance.

Bar elevator conveyor 11.3

This comprises bars carried between a pair of chains, and is often employed in production processes where a single circuit embodies both elevating and conveying. The carrying of parts through dipping and drying operations is a typical application.

The observations made for swing tray elevators apply to this system; it should also be noted that excessive deflection of the bar must be avoided. The limiting factors affecting bar strength are given in Chapter 3.

Plate 76 shows a typical installation, which is now described. This is part of a pre-treatment system for domestic appliances such as washing machine casings and vacuum cleaner parts manufactured from pressed steel. Before stoving or enamelling such items, it is necessary to prepare the metal, which process involves passing through a number of tanks for degreasing and surface treatment. Because of the possibility of the suspended load swinging, smooth running is essential. To achieve this all idler wheels are fitted with ball bearings and two drives are introduced. A guide track is positioned below the two lower wheels, this serving to retain the chain in correct gearing contact with the wheels when the load is carried around the wheel.

Barrel and cask elevator 11.4

In this type a pair of chains is fitted with cradle arms as shown in Fig 11.11. For flat articles, the arrangement is as Fig 11.12.

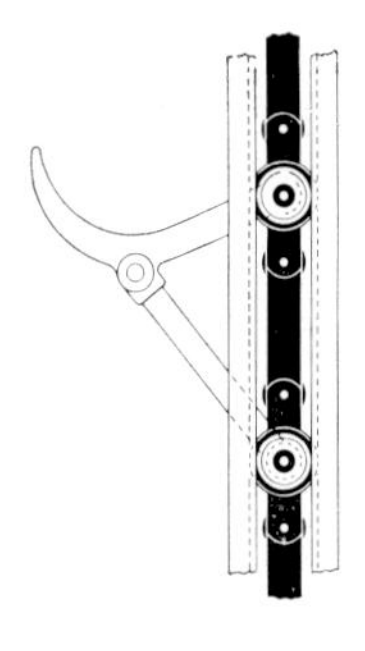

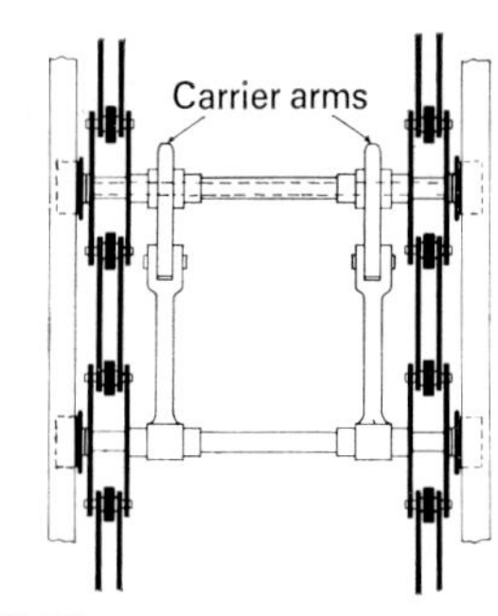

Fig 11.11

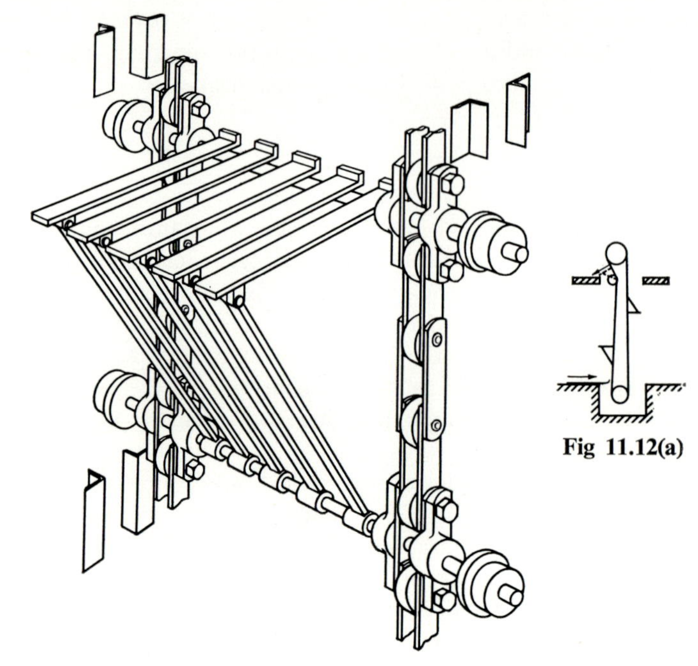

Fig 11.12(a)

Fig 11.12

To prevent the chain toggling under the offset loading conditions continuous guide tracks are essential. In the system shown (Fig 11.12) outboard rollers are used, but guiding on the chain rollers is acceptable for lightly loaded installations within the limits of the roller load for the proposed chain. If the cradle is required to pass around the wheels, the carriers must be able to pivot to accommodate the difference between the linear and chordal pitch. It is usual to operate elevators of this type at speeds of 20 to 50 ft/min. If a snub wheel is fitted as shown in Fig 11.12(a) automatic off-loading is feasible when elevating. Off-loading on the opposite side can be arranged immediately after the barrel has been taken over the top of the headshaft wheel.

Wheel teeth of cast form are suitable for the majority of applications, a minimum of 12 teeth should be satisfactory, but the required diameter of the headwheel must be considered. As a rough guide the diameter should be approximately equal to that of the drum or barrel being handled. Rollers should be mild steel, case-hardened. If the chain is to be guided by outboard rollers, the chain rollers need be only of normal small diameter to gear with the wheel.

CHAIN PULL CALCULATIONS

Chapter 2 will apply, allowance being made for the effect of offset loading as for the finger tray elevator. The friction figure for reaction load assessment would be based on the assumption that rollers are used. The horsepower assessment is also in accordance with Chapter 2, noting that it is the unbalanced load pull which is of significance.

Paternoster elevator 11.5

The principle is illustrated in Fig 11.13 which shows an elevator fitted with trays. The trays are supported by two independent chains at diagonally opposed corners, the attachments usually being extended bearing pins which are free to pivot in bearings attached to the trays.

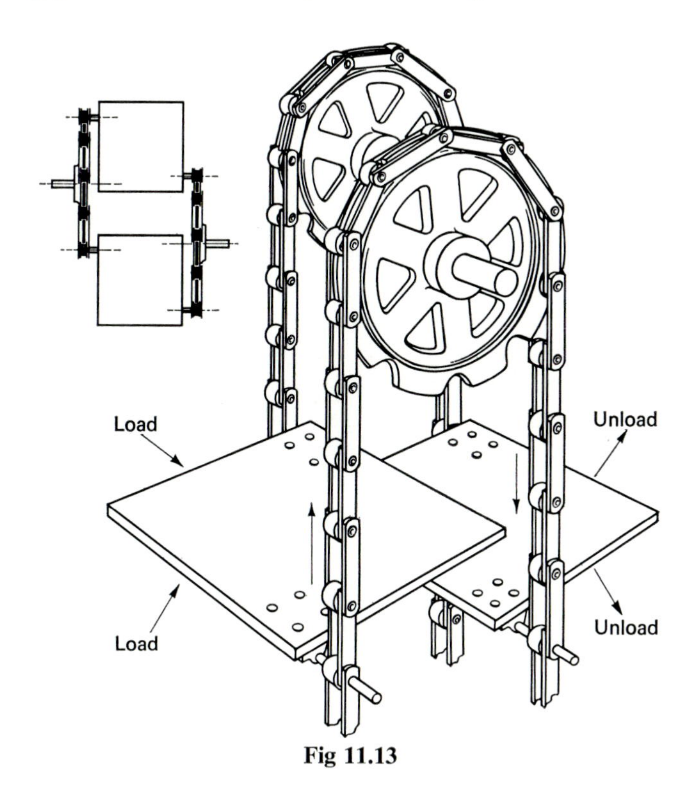

Fig 11.13

An excellent example of this application is the continuous running personnel lift shown in Plate 79 which operates non-stop. The cabs are mounted on the chains as already outlined, thus providing up and down transport side by side. The vertical spaces between the cabs are filled in by board panels, and thus there are no gates or doors required at points of entry and exit, the passengers being confronted either with a blank panel or an open cab. Chain speed is slow so that no difficulty is experienced when entering or leaving the cabs, the cab floors remain horizontal at all times.

Plate 80 show the driving unit and position of the headwheels; the method of obtaining synchronisation of the headwheel speed through gearing will be noted. Plate 81 shows the chain used for a passenger lift, this having extended bearing pins through which a wire rope passes. This rope is a safety feature mentioned in BS 2655:1970. The installation shown runs at a speed of 78 ft/min and has 12 cabs each capable of carrying two people. Six floors are served.

The Paternoster system is not confined to elevating alone, it can also be applied to conveying. Provided the conveyed load is equally distributed about a line drawn diagonally across the trays between the supporting spigots, the trays will remain horizontal. An unevenly loaded tray will tilt, the amount of tilt depending on the clearance between the spigot pins and their bushes.

The chains should be constrained by guide tracks. Paternosters can often be used where the loading and unloading positions preclude the use of a swing tray elevator. The headwheels should have cut teeth in order that close synchronisation is possible. The number of teeth will usually be governed by the tray size, linear spacing and the chain pitch. It is necessary for the headwheels to be mounted on stub shafts with a correctly synchronised drive as typified by Plate 80. Gapped teeth are necessary to clear the spigot pins if these are fitted in the chain plates. Chain adjustment should be provided by downward movement of the tailshaft, the adjustment being securely lockable. In cases where each tailwheel is on a separate stub shaft, extreme care is required to obtain equal adjustment on each of the chains.

CHAIN PULL CALCULATIONS

This is in accordance with Chapter 2.

11.6 Bucket elevator—centrifugal discharge

This system incorporates a series of buckets attached at intervals to one or two chains as shown in Figs 11.14 and 11.15. Material to be moved is fed into the elevator boot by inclined chute, the buckets then picking it up by a scooping or dredging motion. Discharge relies on centrifugal action to throw the material clear of the preceding bucket as the headwheel is negotiated. The elevator may be vertical or slightly inclined from the vertical the former taking up a relatively small space and having a simpler structure.

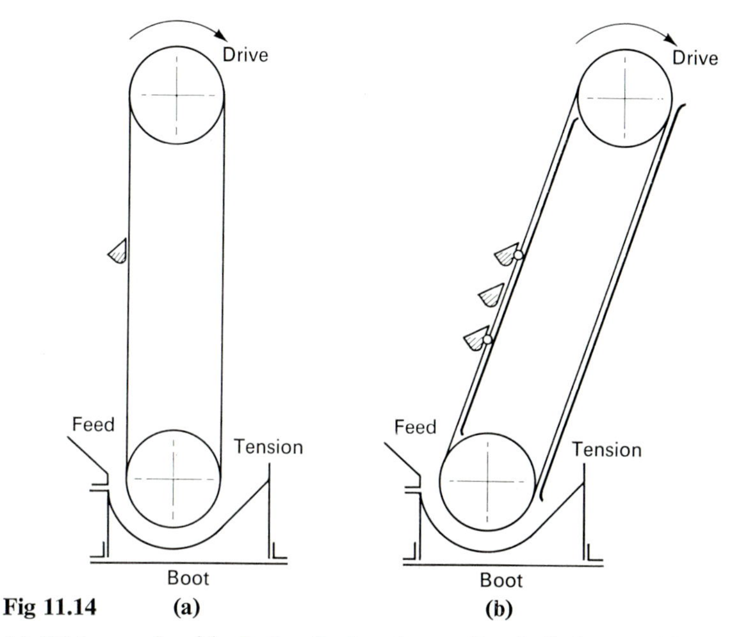

Fig 11.14 **(a)** **(b)**

(a) High speed, with dredge feed and centrifugal discharge **(b)** Medium speed, with dredge feed and centrifugal discharge

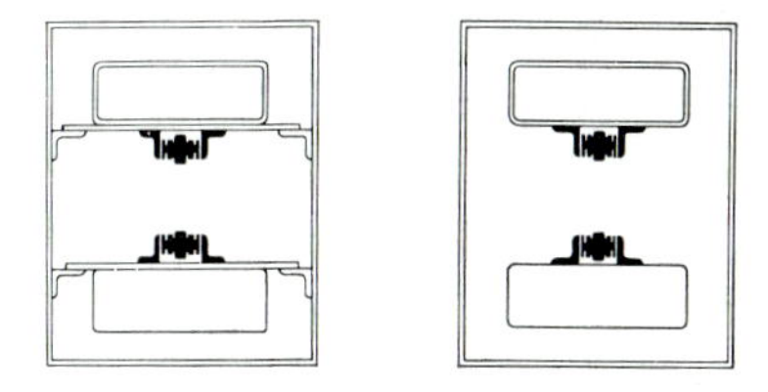

Fig 11.15

This type is particularly useful for handling materials not exceeding 3 in cube. Materials having abrasive characteristics can be dealt with, but a high wear rate of buckets and chain must then be accepted. Versions with both single and double strand chain are commonly used, the selection of the latter type depending on the width of bucket required. Two chain strands are necessary if the bucket width is 16 in or more. It is usual to operate elevators of this type at a chain speed of about 200 ft to 300 ft/min, but each application must be considered individually in relation to achieving an effective discharge, the latter being dependent on the peripheral speed of the bucket around the headwheel. Other important factors influencing discharge are the type of material, bucket shape and spacing.

Feed chute angles vary with the materials handled but are generally arranged at 45° to the horizontal. Material should be fed to the buckets at or above the horizontal line through the boot wheel shaft. Where bucket elevators are an integral part of a production process, it is usual to have interlocks on the conveyor and elevator systems to avoid unrestricted feed to any unit which may for some reason have stopped.

The selection of the correct shape and spacing of the buckets relative to the material handled, are important factors in efficient operation. Spacing of buckets depends upon the type of bucket and material handled, but generally 2 to $2\frac{1}{2}$ times the bucket projection is satisfactory. Bucket capacities as stated by manufacturers are normally based on the bucket being full, but this capacity should be reduced in practice to about 66% to ensure that the desired throughput is obtained.

Solid bearing pin chain is essential for other than light, clean duty applications. Chain pitch is normally dictated by bucket proportions and desired spacing. Mild steel case-hardened rollers should be used but where these are not required for guiding purposes, small diameter gearing rollers may be used.

Due to the high loadings which can occur during dredging, particular care is necessary in ensuring that the chain attachments, buckets and bucket bolts are sufficiently robust to withstand these loadings. Normally K2 attachments are used; Figs 11.14 and 11.15 illustrate typical examples. This means that lower chain speeds can be used to effect adequate material discharge speeds, as the buckets operate at a greater radius than the wheel pitch circle diameter.

The selection of the headwheel pitch circle diameter is related to obtaining correct discharge as described later. Generally the headwheel should have a minimum of 12 teeth, otherwise the large variation in polygonal action which occurs with fewer numbers of teeth will cause irregular discharge and impulsive loading. This will result in increased chain pull, greater chain wear and stress on the buckets. Where the material handled has abrasive characteristics and/or high tooth loadings exist, steel wheels are necessary. For extremely high engaging pressures the wheels should have flame-hardened teeth.

To aid bucket filling the boot wheel size should be the same as that of the headwheel. Where abrasive materials are involved boot wheels should be manufactured from steel. Irrespective of size or material handled the boot wheel teeth should be side relieved to reduce material packing between the tooth root and the chain.

Chain adjustment is normally provided by downward movement of the boot shaft, and allowance should be made for this in the boot design. Certain materials handled by this type of elevator have a tendency to pack hard, and therefore material in the boot should be cleared manually before adjusting the chains to avoid fouling.

On long centre distance installations, guiding of the chain is necessary to avoid a whipping action which can be promoted by the dredging action. It is not always necessary to provide continuous guide tracks, and common practice on say a 70 ft elevator would be to introduce three equally spaced 6 ft lengths of guide for each strand of chain. Inclined elevators must have continuous chain guides irrespective of length of elevator.

The discharge operation of a centrifugal elevator is shown in Fig 11.16, the sequence being as follows:—

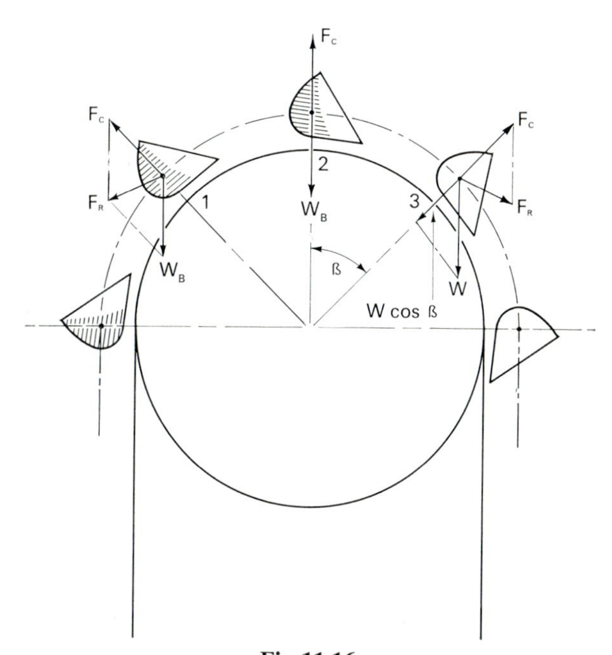

Fig 11.16

The weight of material in the bucket acts downwards due to gravity whilst centrifugal force tends to throw the material radially outwards; therefore at position (1) the resultant force holds the material in the bucket. In position (2) the forces directly oppose each other. If these are equal the resultant force is zero and immediately the bucket passes this point material will be discharged from the bucket.

Discharge at this point usually takes place with free flowing materials such as grain and flour. By considering the forces shown in Fig 11.16 the bucket speed required to effect discharge at this position can be calculated as follows:—

$$\text{Centrifugal force} = \frac{W_B v_M^2}{g r_M} \text{ (lbf)}$$

Where W_B = Weight of material in bucket (lb)
v_M = Velocity of material at the centre of gravity (ft/sec)
g = Acceleration due to gravity (32·2 ft/sec²)
r_M = Radius of centre of gravity of material (ft)

For equilibrium to occur at position (2) then:—

$$W_B = \frac{W_B v_M^2}{g r_M} \text{ or } v_M^2 = g r_M$$

For heavy coarse materials such as coal, bucket speeds are generally lower to reduce the effect of centrifugal force, this normally being reduced to 70% of W_B. As a result the material will still be held in the bucket and discharge will not occur until position (3) is reached, when the centrifugal force balances the radial component of the weight W_B. The ideal speed for an elevator handling a material such as coal can be calculated as follows:—

Weight of material in bucket = W_B (lb) and if centrifugal force is to be limited to 0·7 W_B we have:—

$$0{\cdot}7\ W_B = \frac{W_B v_M^2}{g r_M} \text{ or } v_M^2 = 0{\cdot}7\ g r_M$$

It is now possible to determine the angle β which is the angle at which the bucket will discharge. For materials such as coal this is normally about 40°.

$$W \cos \beta = \frac{W_B}{g r_M} v_M^2 \text{ or } \cos \beta = \frac{v_M^2}{g r_M}$$

When material leaves the bucket the tendency is for it to travel straight on at the projected velocity. It is, however, acted upon by gravity, and thus the material follows a downward trajectory. The vertical displacement of the material is given by:—

$$\tfrac{1}{2} g t^2 \text{ (ft)}$$

Where g = Acceleration due to gravity (32·2 ft/sec²)
t = Time lapse after leaving the bucket (secs)

If the initial material velocity is v_M ft/sec the horizontal displacement in ft will be given by the product of $v_M t$. By relating the two factors the material path can be traced and the correct position for the delivery chute determined. This also enables a check to be made on the spacing of the buckets to ensure that the material discharged from one bucket does not foul the back of the preceding bucket. As an example, if the projected velocity is taken as 240 ft/min, the horizontal distance travelled in $\frac{1}{2}$ sec $= \frac{240}{60} \times 0{\cdot}5 = 2$ ft and the vertical displacement will equal

$$\tfrac{1}{2} \times 32{\cdot}2 \times (\tfrac{1}{2})^2 = 4 \text{ ft (approx)}$$

The graph Fig 11.17 shows a comparison at a projected velocity of 4 and 10 ft/sec.

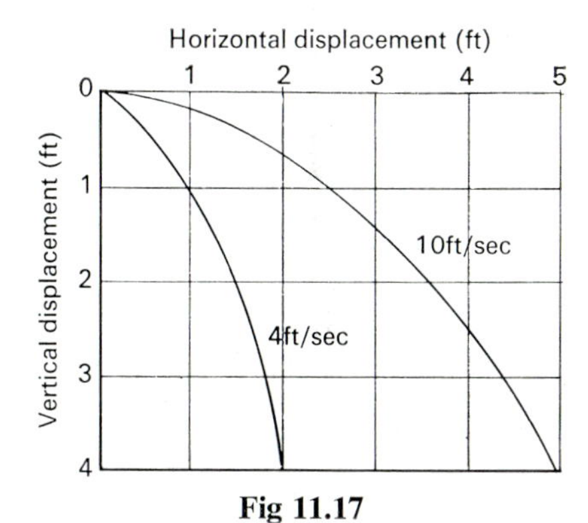

Fig 11.17

CHAIN PULL CALCULATIONS

This is in accordance with Chapter 2 but a further load pull will be induced, caused by the dredging action and must be added to the value derived from the formula. Thus:—

$$\text{Dredging pull} = \frac{360 \times W_B}{p_s} \text{ (lbf)}$$

Where W_B = Weight of material in each bucket (lb)
p_s = Spacing of buckets (in)

The foregoing applies to both vertical and inclined elevators. Horsepower assessment is in line with Chapter 2 noting that it is the unbalanced load that is significant, also the additional load due to dredging action has to be catered for. A useful approximate figure for motor horsepower for vertical elevators can readily be obtained from the following:—

$$\text{hp} = \frac{\text{tph} \times \text{lift (ft)}}{442}$$

Where tph = Tons/hr of material elevated

The following example of selection is given as a guide.

EXAMPLE

A centrifugal discharge spaced bucket elevator is required to handle 16 tons of coal per hour. Chain wheel centres are 100 ft and weight of material is 55 lb/cu ft.

A certain amount of trial and error will be necessary when selecting chain sizes and bucket capacity, and to assist selection the chart shown in Fig 11.18 is useful.

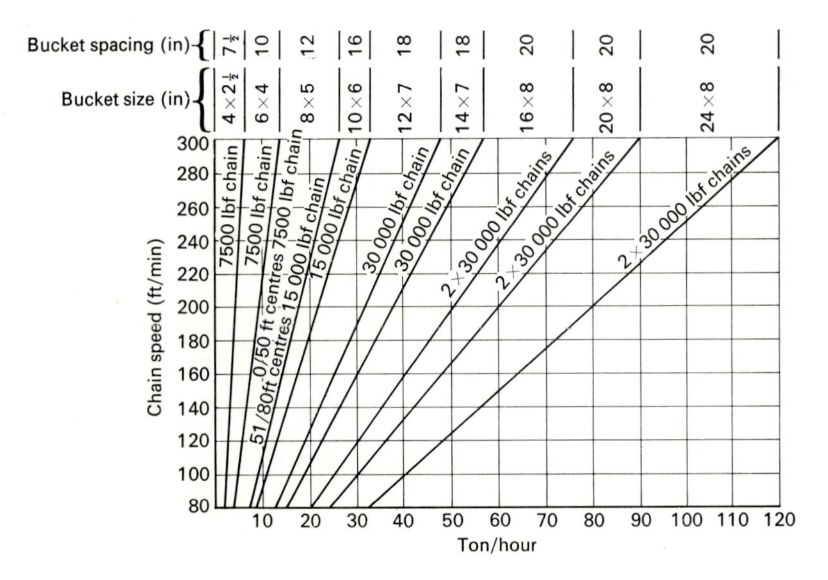

Maximum centres: 80 ft. Material: 50 lb/ft³
For lighter and heavier materials due allowance must be made when making selection

Fig 11.18 Spaced bucket elevator

This shows that a bucket 8 in long by 5 in projection appears, adequate. The chain provisionally selected is a 15,000 lbf 6 in pitch type, and assuming that the headwheel has 12 teeth with PCD of 23·18 in, the radius (r_M) at the centre of gravity of the material is determined as shown in Fig 11.19.

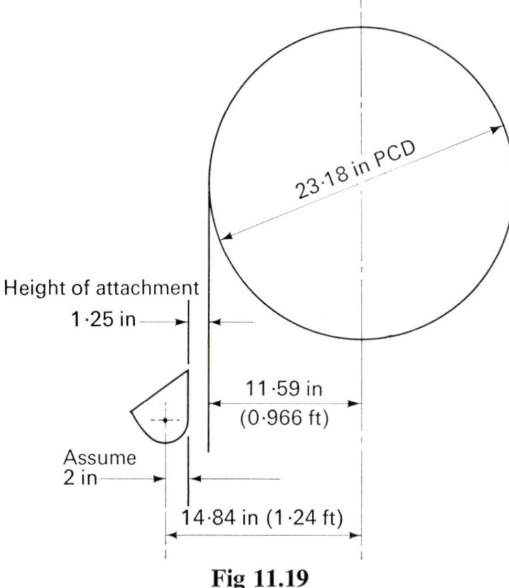

Fig 11.19

It is now possible to determine the bucket speed around the head-wheel in order that discharge will be effective.

$$v_M^2 = 0{\cdot}7\ gr_M$$
$$v_M = \sqrt{0{\cdot}7 \times 32{\cdot}2 \times 1{\cdot}24} = 5{\cdot}286 \text{ ft/sec}$$
$$= 317 \text{ ft/min}$$

$$\text{Speed of chain} = \frac{317}{1{\cdot}24} \times 0{\cdot}966 = 247 \text{ ft/min}$$

A check with the chart indicates that the selection is proceeding satisfactorily. The bucket capacity can now be established.

$$\text{Weight of material in bucket} = \frac{C \times p_S}{S \times 12} \text{ lb}$$

where C = Elevator capacity (lb/min)
p_S = Pitch of buckets (in)
S = Chain speed (ft/min)

$$\text{Weight of material in bucket} = \frac{16 \times 2{,}240 \times 12}{60 \times 247 \times 12} = 2{\cdot}417 \text{ lb}$$
$$= 0{\cdot}0439 \text{ cu ft}$$

As the bucket should only be $\frac{2}{3}$ full the actual bucket capacity should be 0·066 cu ft. From the manufacturer's catalogue the nearest suitable bucket has a capacity of 0·07 cu ft. This is an 8 in long bucket weighing 4·4 lb. If a bucket of suitable capacity is not catalogued, the calculations will have to recommence by using a bucket with larger or smaller projections as required. The chain pull can now be calculated as follows:—

Number of buckets on loaded strand = 100
Weight of material on loaded strand = $100 \times 2{\cdot}417 = 242$ lb
Weight of buckets = $100 \times 4{\cdot}4 = 440$ lb
Pull due to dredging = $\frac{360 \times 2{\cdot}417}{12} = 73$ lbf
Weight of chain = $100 \times 3{\cdot}42 = 342$ lb
Weight of attachments = $100 \times 0{\cdot}4 = 40$ lb
(K attachments both sides every outer pitch)
Total chain pull = 1,137 lbf

Therefore the original selection of 15,000 lbf chain would have a factor of safety of 13 which is satisfactory.

$$\text{Horsepower required} = \frac{(73+242) \times 247}{33{,}000} = \frac{315 \times 247}{33{,}000} = 2{\cdot}37$$

11.7 Spaced bucket positive discharge elevator

This type of elevator has a series of buckets fixed at intervals between a pair of chains. The material is picked up by scooping or dredging as in the centrifugal discharge elevator, but since a comparatively low speed is used then bucket filling is more effective. The chains are mounted outboard of the buckets as shown in Fig 11.20. This facilitates the discharge operation which is accomplished by inverting the bucket, the contents of which fall clear of the preceding bucket.

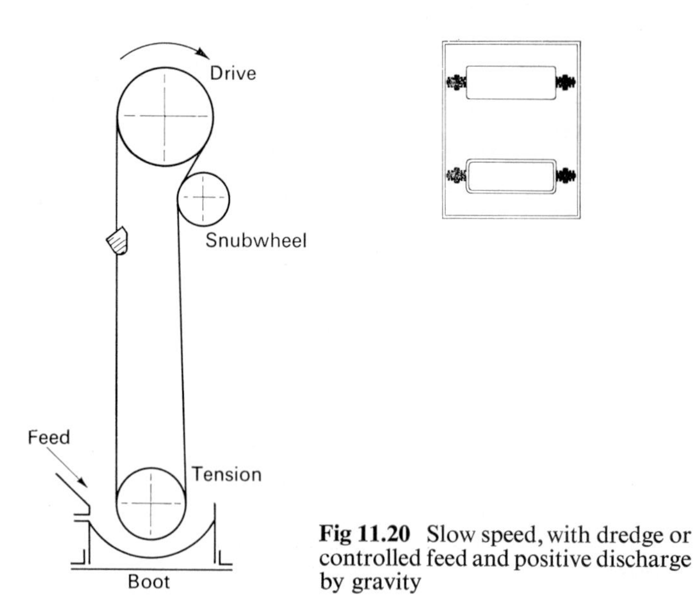

Fig 11.20 Slow speed, with dredge or controlled feed and positive discharge by gravity

It will be appreciated that there is no critical relationship between operational speed and effective discharge. The speed should be kept as low as possible, in line with obtaining a satisfactory discharge and throughput. As a result the overall wear rate on all components can be kept low. Selection of the correct shape and spacing of buckets has an important influence on efficient operation. Generally, the design of bucket used for materials normally handled has a depth equal to or slightly less than the projection, while the linear spacing should be approximately twice that of the bucket projection. When filling by dredging action, a fill of about 75–80% of the bucket volume can be assumed.

The size of the boot wheel is governed by the selected size of headwheel, along with the snub wheel position and the fact that it is normal practice for the loaded and unloaded chain strands to be parallel, as shown in Fig 11.21.

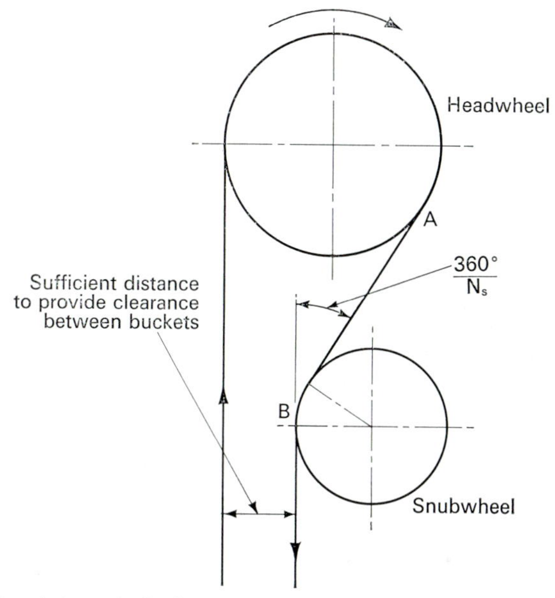

Fig 11.21

Because of their particular function, and considerations of space, snub wheels are smaller than the head and boot wheels. On high capacity elevators they are subjected to impact tooth loading, caused by the reaction load of the hanging chain and buckets as they descend after negotiating the headwheel. This tooth impact will be greatest at the bucket positions, and therefore if the latter are spaced at an even number of chain pitches the snub wheel should have an uneven number of teeth to equalise wear. Conversely, if the buckets are unevenly spaced relative to chain pitch, an even or odd number of wheel teeth can be used providing the selected number is not divisible by the bucket spacing pitch.

The recommended snub wheel position is shown in Fig 11.21, the distance A–B being the equivalent of an equal number of pitches plus half a pitch; normally 4·5 are satisfactory. Snub wheels should be of steel with flame-hardened teeth.

The chart shown in Fig 11.18 should be referred to, for determining the bucket proportions, spacings and chain size. Other calculations and information can be obtained from the section dealing with centrifugal discharge elevators.

11.8 Continuous bucket elevator

DESCRIPTION AND CHAIN TYPE

In this arrangement of elevator, the buckets are fitted to the chain or pair of chains in close proximity, forming virtually an endless sequence, the buckets being so designed that when discharge takes place over the headwheel the back of one bucket functions as a chute for the following one. There are broadly two types of bucket in general use. Plate 86 shows the overlapping type which reduces spillage to a practicable minimum. The second type, Plate 87, does not overlap but has the minimum gap between the bucket edges, which curtails the spillage, particularly at the on-loading position.

Bucket design will vary with the capacity required and material handled. It is usual for the buckets to be fixed above the chain plates as shown in Fig 11.23(a) and (b), but if for higher capacity, fitting between two chains as in Fig 11.23(c) is necessary. This latter arrangement is referred to as a 'Supercapacity' continuous bucket elevator. The nominal bucket depth is equal to, or a multiple of, the chain pitch. The in-feed is usually by way of an inclined chute, the position on a vertical elevator being normally two to three bucket spacings above the tailshaft. In that case, if spillage should occur, the material will be automatically accepted by the following buckets, minimising build-up of material in the boot. If the elevator is inclined slightly, both the feed and discharge are assisted, and in such a case the in-feed may be located at a position one or two bucket spacings upwards on the loaded side of the chain.

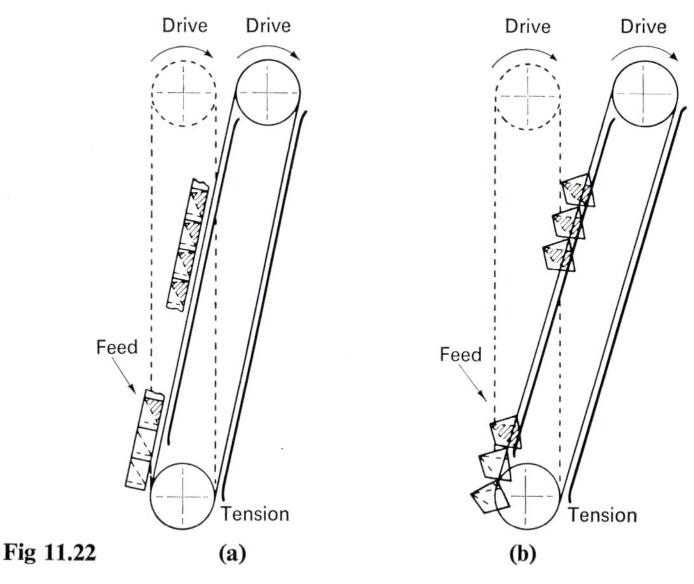

Fig 11.22 **(a)** **(b)**

(a) Vertical elevator **(b)** Inclined elevator

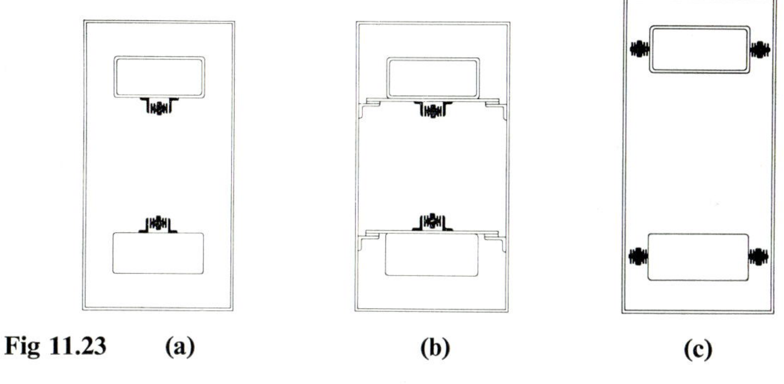

Fig 11.23 **(a)** **(b)** **(c)**

(a) Vertical bucket elevator, K attachments **(b)** Inclined elevator, K attachments **(c)** Super capacity inclined/vertical elevator, G attachments

As the dredging action in the boot is minimised, the continuous bucket design is particularly suitable for handling abrasive bulk material. Chain speed is normally about 100 ft to 175 ft/min and as the conveying effect is continuous a high throughput can be achieved.

For information on the selection of chains, chain wheels and guiding etc., reference should be made to the section dealing with centrifugal discharge elevators and Fig 11.24.

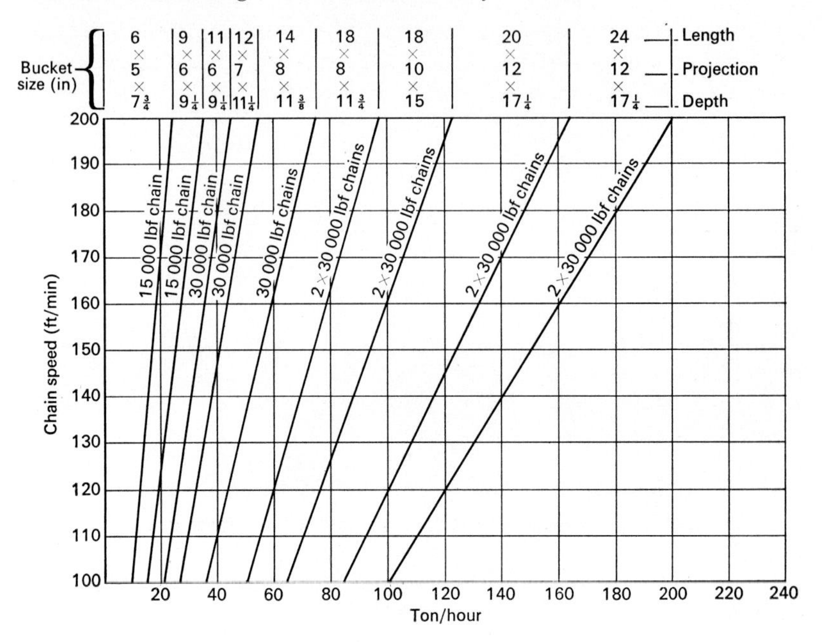

Maximum centres: 80 ft. Material: 50 lb/ft³
Solution irrespective of centres up to 80 ft
For lighter and heavier materials due allowance must be made when making selection.

Fig 11.24 Continuous bucket elevator

Three common methods of attaching the buckets to the chains are illustrated in Fig 11.23 but there are many other feasible methods employing standard K and G attachments.

CHAIN PULL CALCULATIONS

For this, reference should be made to Chapter 2. If there is dredging action as just described, the additional load pull must be added to the value derived from the formula. The additional load pull is obtained from

$$\text{Dredging pull (lbf)} = \frac{144 \times W_B}{p_S}$$

Where W_B = Weight of material in each bucket (lb)

p_S = Spacing of buckets (in)

The foregoing applies both to vertical and inclined conveyor elevators.

The horsepower to drive is calculated in accordance with Chapter 2 noting that the unbalanced load is significant, if dredging action is involved this must also be covered.

Gravity bucket elevator **11.9**

DESCRIPTION AND CHAIN TYPE

This incorporates buckets which are free to pivot about attachment points between a pair of chains, in such a way that bulk materials may be carried in one plane and at different elevations as required. At all times the buckets remain horizontal. A typical layout is shown in Fig 11.25.

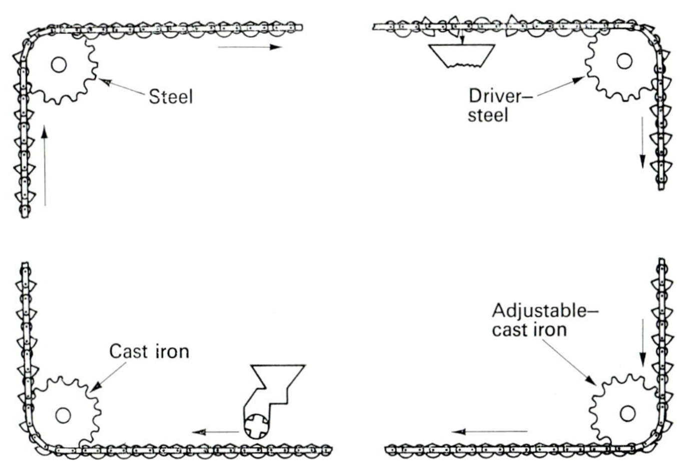

Fig 11.25

Loading of the individual buckets is normally done by means of a controlled feeding device such as a star feeder. However, the buckets may be made to overlap, in which case a continuous feed can be arranged. Discharge is effected by fitting the buckets with trip arms or cast quadrants engaging appropriate devices on the elevator structure as shown in Plate 89. This can be done at different points along a horizontal run, or at a wheel position as illustrated in Fig 11.26. In this manner, different materials may be handled at the same time by suitable phasing of the feed and discharge positions appropriate to a particular material.

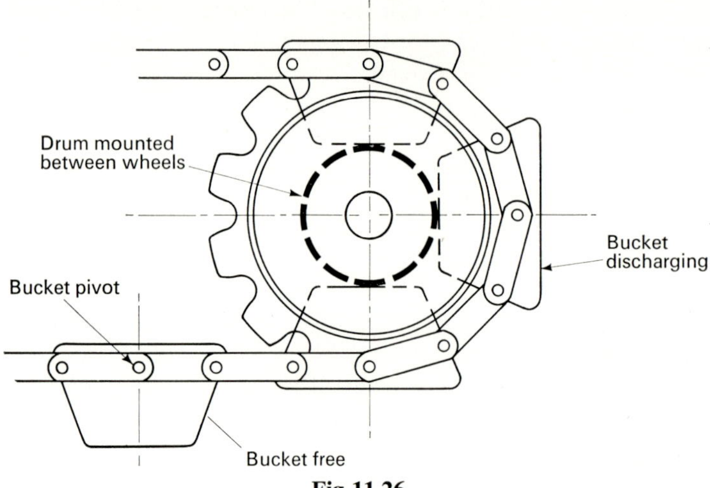

Fig 11.26

This type is particularly suitable for handling large volumes of bulk materials which are abrasive, corrosive or with combinations of such characteristics. The buckets can be obtained with resistant linings. The method of controlled feed reduces contact between the chain and the material, with obvious benefit to the former. Typical materials handled are coal, ash, clinker and cement, amongst many others.

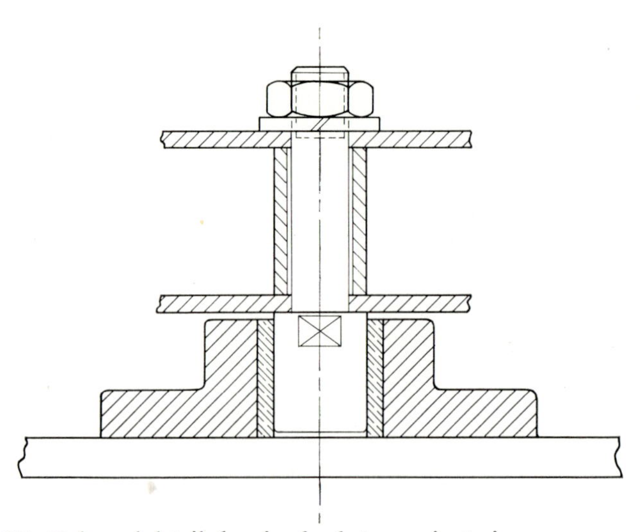

Fig 11.27 Enlarged detail showing bucket on spigot pin

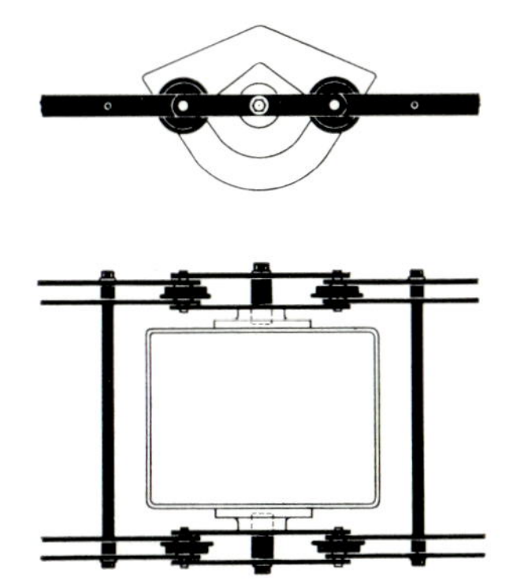

Fig 11.28 Spigot pins and staybars bolted through chain plates

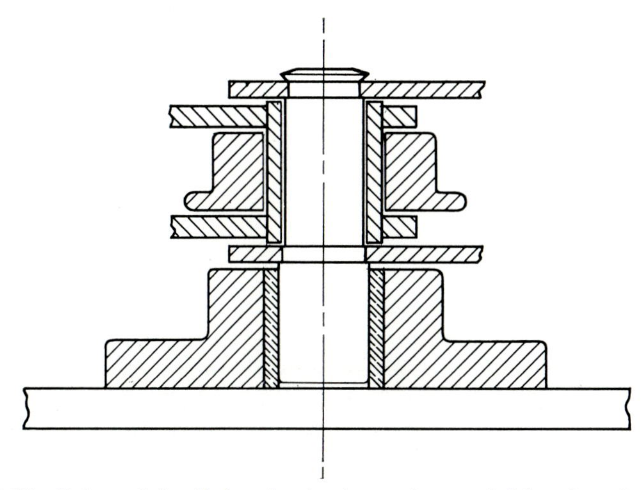

Fig 11.29 Enlarged detail showing bucket and extended bearing pin

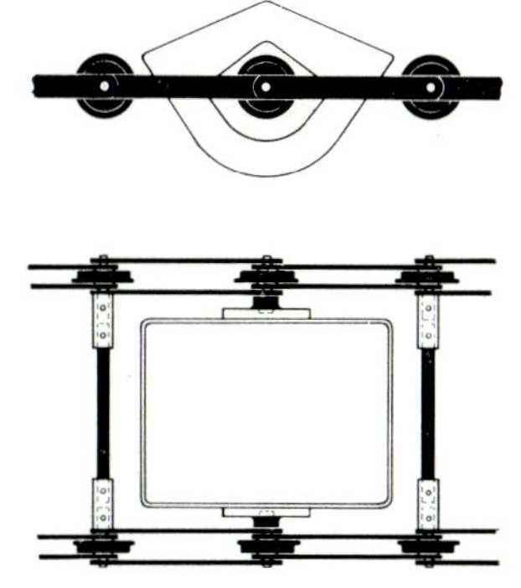

Fig 11.30 Spigot pins and special staybars bolted through hollow bearing pins (extended solid bearing pins alternative to spigot pins)

Chain pitch is normally dictated by the proportions of buckets. Flanged rollers are to be preferred, but plain rollers in conjunction with guide strips are acceptable. Typical examples of attachment methods are shown in Figs 11.27, 11.28, 11.29 and 11.30, these being more commonly identified with heavy duty applications. The staybars ensure chain parallelism and maintenance of transverse centres, and also that the buckets remain free to pivot at all times. For light duty systems, spigot pins bolted through hollow bearing pins are quite satisfactory. On large installations the drivers and top wheels should be made of steel while cast iron is satisfactory for the lower wheels. Drive wheels should have a minimum of 12 teeth. Idlers having 8 teeth are in order, but clearance between buckets and through shafts may call for larger wheels. The teeth must be gapped if necessary to clear staybars or similar fittings. In order to maintain correct chain adjustment, the most convenient wheel after the driver may be selected as the adjusting medium. It is normal to employ a through headshaft. The circuit shown in Fig 11.25 indicates a chain lap on the driving wheel of only 90°. The maximum possible number of teeth must therefore be used, and certainly not less than 12.

CHAIN PULL CALCULATIONS

The method stated in Chapter 2 is applied, using the section method. Referring to Fig 11.25 as an example, the calculation would be commenced at a position immediately after the drive wheel. Horsepower assessment is obtained in accordance with Chapter 2. Note that the downward leg at the right hand side of the circuit (Fig 11.25) would assist the drive.

Plate 70 Newsprint reel lowerator

To lower reels of newsprint, weighing 1,400 lb and measuring 3 ft diameter × 5 ft long, from ground floor receiving bay to storage basement, a swing tray lowerator is used. The trays are shaped cradles carried from spigot pins and made from rolled steel channel faced with formed hardwood blocks. The loading point is shown where the reels are loaded by rolling onto the cradles, the padded stock block being shown in the rear. At the off-loading level the reel comes into contact with fixed, forward inclined wooden blocks spaced between the cradle blocks. These cause it to roll forward, off the cradle.

Plate 71 Bread cooler

One of the most firmly established forms of swing tray conveyor is the bread cooler. These installations are of considerable length, but to deal with the enormous quantities of loaves baked daily in a modern bakery it is still necessary to make trays of considerable lateral width and to run the pair of continuous chains on several tiers to give maximum cooling time within as small a space as possible. Since the load is freely suspended and speed is slow it is essential to ensure smooth running of the chains.

Plate 72 Pottery drier

This pottery drier utilises two conveyor chains matched to run as a pair, with swing trays 20 ft long carrying a total load of 40 tons in horizontal and vertical paths through the drier. The trays are supported on spigot pins bolted through hollow bearing pins of the chains. The conveyor is driven at four points by chain drives from a single motor; it is driven on both sides to prevent twist in the long shafts.

Plate 73 Pottery drier for domestic cups

The circuit for this drier embraces both horizontal and vertical runs. The cups are carried in containers mounted both sides of a single chain, the containers being free to pivot.

Plate 74 Finger tray lowerator for newspapers

This is an example of a finger tray installation carrying newspapers from an upper floor to a ground floor loading bay, the capacity being 150,000 newspapers per day. The photograph shows the upper floor loading hatch. Newspaper bundles are placed on a slatted platform with slatted back fence from whence they are lifted by the 'combing' action of the rising finger tray which is shown, with its triangular bracket, just about to rise through the platform. The newspapers are then carried over the head wheel and descend on the other side, a finger tray being visible behind the wire mesh guard. At the lower end, the finger tray passes through a similar slatted platform and the newspapers are 'combed' off.

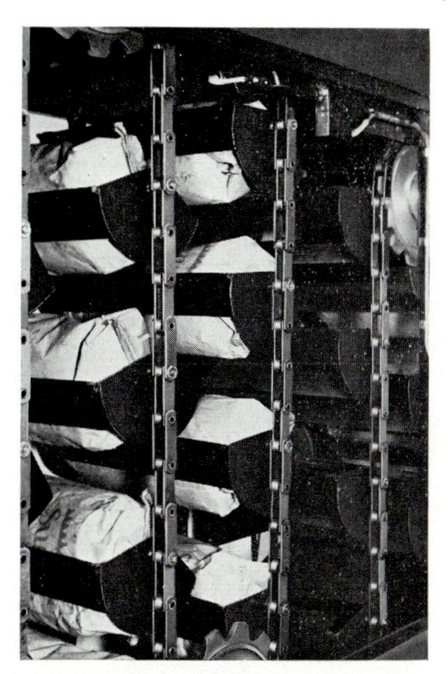

Plate 75 Automatic coal vending machine

Capacity:—49 bags each 28 lb coal or 14 lb smokeless fuel. Swing trays are carried on special spigot pins bolted through hollow bearing pins at every fourth pitch point.

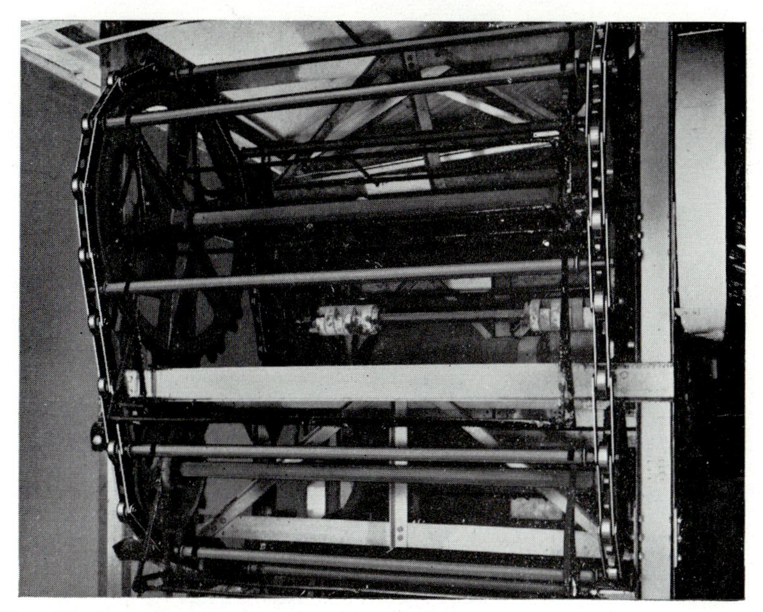

Plate 76 Bar elevator-conveyor

Plate 77 Barrel elevator

Fixed arm elevator for barrels, casks or drums. 5 in pitch, hollow bearing pin chain. The cradles run in guides with rollers and are secured to special sideplate in chains.

Plate 78 Sling or pocket elevator

This combined elevator and conveyor consists of two strands of roller chain with staybars at intervals to support the slings. As the load passes from one plane to another it rolls into a new position in the sling but is not discharged. The size of sling is related to the maximum size of package or goods to be handled.

Plate 79 Paternoster lift

Plate 80 Paternoster lift—driving unit and headwheels

Plate 81 Paternoster lift—headwheel with safety rope

Plate 82 Clinker elevator

A view of spaced buckets passing round the headshaft wheels of a clinker elevator illustrates how angles riveted to the sides of the buckets may be employed to bolt the buckets to K2 attachments on the chains. The dredging action in the boot imposes additional bucket loading and they must be of robust construction as this photograph, showing the internal stay reveals.

Plate 83 Grain elevator

Wet grain at the rate of 90 sacks per hour is handled by this single strand elevator.

Plate 84 Sand elevator

A section of an inclined spaced bucket elevator for wet sand is shown to illustrate the manner in which buckets may be attached to a single chain, one of the buckets having been removed to show the K2 attachments.

A skidder bar bolted between bucket and attachment slides on rolled steel angles at the sides of the conveyor which serve to support the bucket. Since the sand is wet as dredged in the boot, the buckets are perforated for drainage purposes. A high safety factor is applied in selecting the chain to afford increased bearing area on wearing parts in view of the highly abrasive nature of material handled. Use of a dry lubricant is recommended.

Plate 85 Heavy duty spaced bucket elevators for chemicals

A pair of 85,000 lbf breaking load chains with bent over K attachments integral with the sideplates on one side of each chain carry spaced buckets handling sulphate of ammonia and potash at the rate of 120 tons per hour. The temperature of the material is 115°C.

Plate 86 Continuous overlapping bucket elevator for gravel

This bucket arrangement is useful where abrasive materials such as gravel have to be handled, since contact between material and chain is limited. The buckets span two pitches of chain, being bolted to K2 attachments on the outer link on one side of each pair of chains. Not only the side but the rear plate of each bucket overlaps the following bucket.

Plate 87 Continuous inclined bucket elevator for gravel

The buckets, whilst being bolted to K2 attachments on a pair of chains, do not overlap but rely on fine clearances to avoid spillage. In this case, the chains are located inboard beneath the buckets and the attachments are fitted on both sides of every outer link. The chains act solely as traction media, each bucket having a flanged outboard roller at each side to guide and to carry the load, and provide support on the return run.

Plate 88 Continuous bucket elevator handling peas

This small elevator handling peas in a canning factory, shows the versatility of elevators and illustrates how the back of one bucket serves as the chute down which the following bucket discharges. The bucket design in this case is vee-shaped, pivoting on staybars fitted across the pair of chains at every pitch point.

Plate 89 Gravity bucket conveyor handling cement clinker

The installation employs 85,000 lbf chain with grease gun lubrication. Staybars between the buckets join the chains and the flange rollers are hardened to withstand the highly abrasive and dusty conditions. The trip arms mounted at the sides of the buckets vary in their settings to trip alternate buckets only at their particular location. By this means, two discharge points, each of equal quantity, are arranged.

Plate 90 Gravity bucket elevator for swarf

This installation conveys metal swarf to a hopper over a centrifugal oil extractor which separates oil from the swarf. The free swinging buckets are carried between a pair of chains by staybars passed through hollow bearing pins, these also space the chains.
Bucket discharge is accomplished by mounting a drum between the pair of wheels over the hopper, the drum diameter being such that it contacts the base of the bucket at that point and inhibits the free swing so that the bucket discharges. After passing the drum, the bucket is free again and rights itself by its own weight. A restraining plate, suitably positioned, prevents excessive swing.

12

Adapted Transmission and Special Purpose Conveyor Chains

12.1 Adapted transmission chains

In some conveying systems, considerations of size, speed or other factors may preclude the use of normal type of conveyor chain. These systems often come within the scope of standard power transmission chains with attachments and these are then designated Adapted Transmission Chains. Standard variations of chain and attachments are shown in Figs 12.1 to 12.4. Transmission chain without attachments can be used alone to convey articles resting on plate edges. Some of these chains and attachments are available in stainless steel.

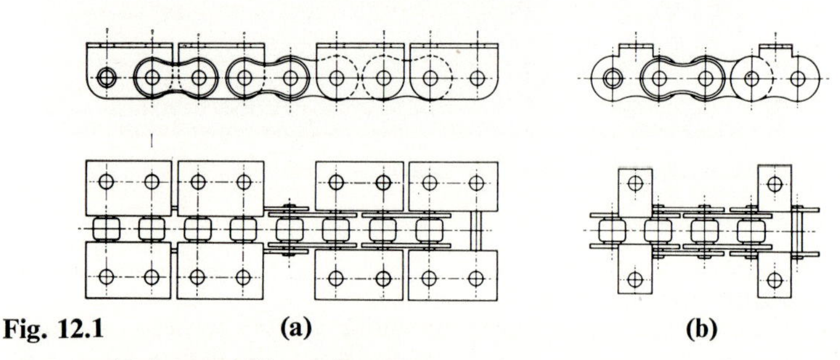

Fig. 12.1 **(a)** **(b)**

Chains with integral K attachments:
(a) Long K2 attachments. **(b)** Short K1 attachments

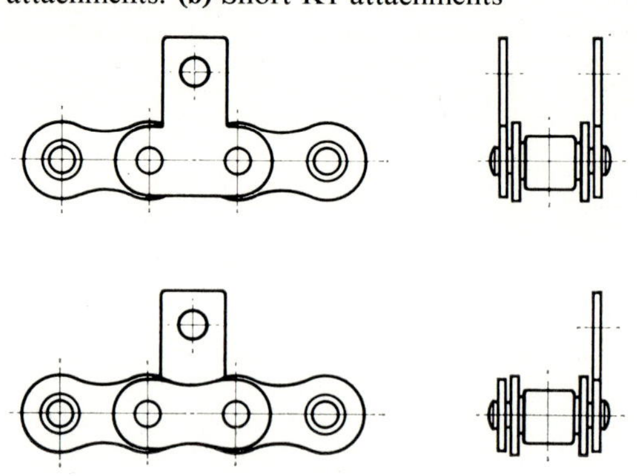

Fig 12.2 Chains with M attachments

Fig 12.3(a) **(b)** **(c)** **(d)**

Chains with extended pins. **(a)** Outer link with one extended pin **(b)** Connecting link with one extended pin **(c)** Outerlink with two exteneded pins **(d)** Connecting link with two extended pins

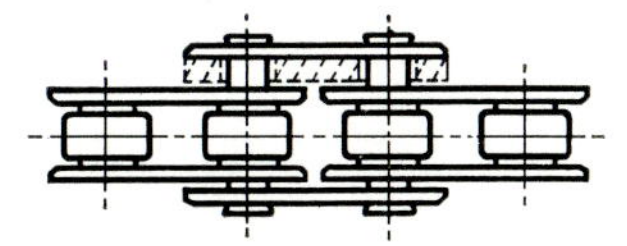

Narrow 0·5 in pitch chain with outer link from wide 0·5 in pitch chain

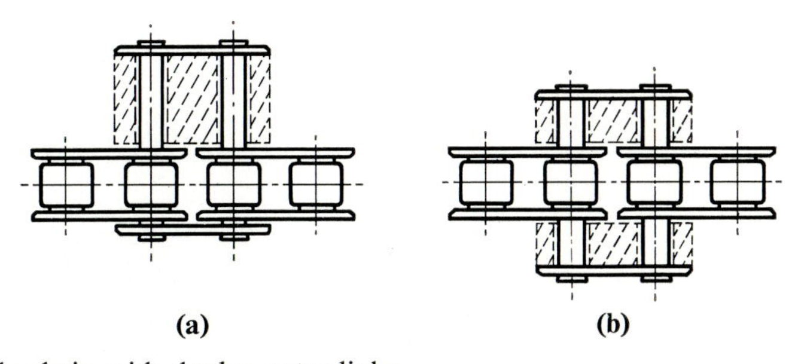

(a) **(b)**

Simple chain with duplex outer links
(a) Attachment one side **(b)** Attachment both sides

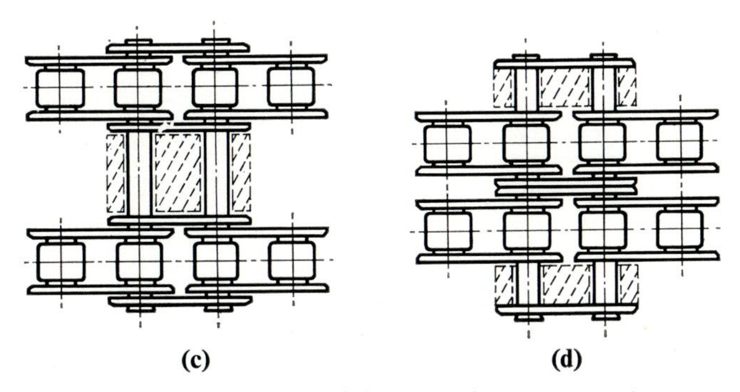

(c) **(d)**

c) Twin simple chains with triplex joint. Attachments central
(d) Duplex chain with triplex joint. Attachments both sides

Fig 12.4

Typical applications are shown in Plates 91 and 92, where a tin-printing drying oven incorporates two strands of 1·0 in pitch transmission chain which have the bearing pins extended on one side at each pitch. The frames carrying the sheets of metal after printing are supported on the extended pins, the roots of the carrier frames being holed to locate on the extended pins at each pitch point, and secured by split pins.

In recent years increased oven output coupled with a reduction in sheet thickness have necessitated the use of no back bend tinprinter chain. These chains as shown in Figs 12.5 and 12.6 incorporate large outboard rollers which with their lower rolling friction promote smooth running which is essential when handling the sheets.

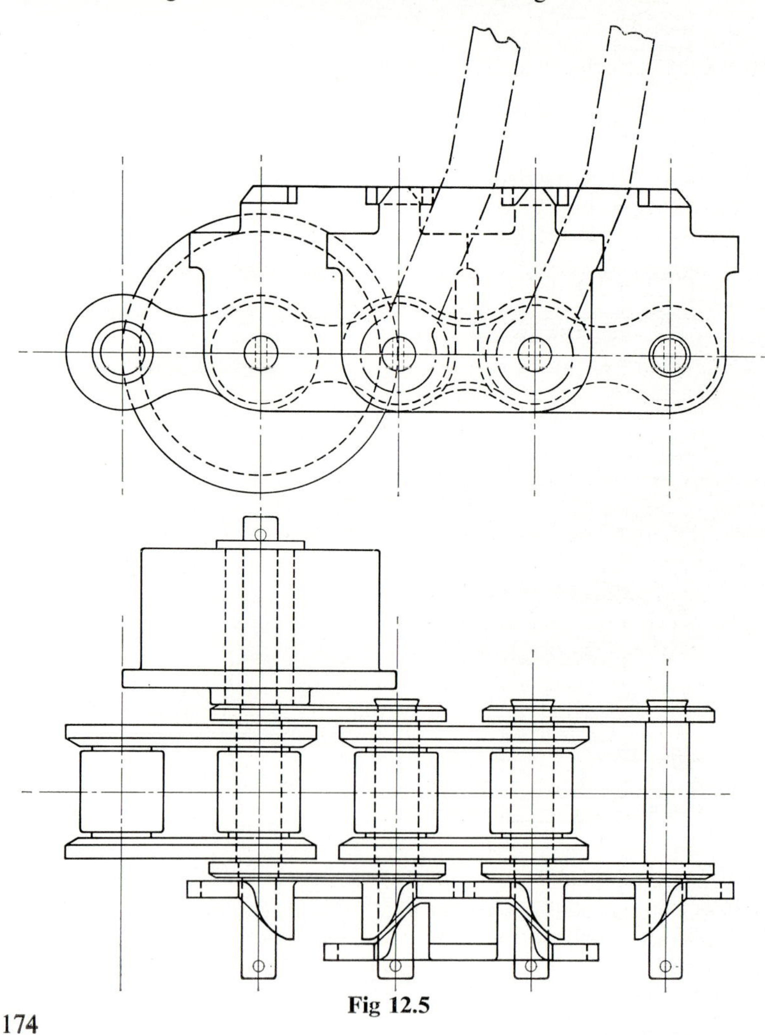

Fig 12.5

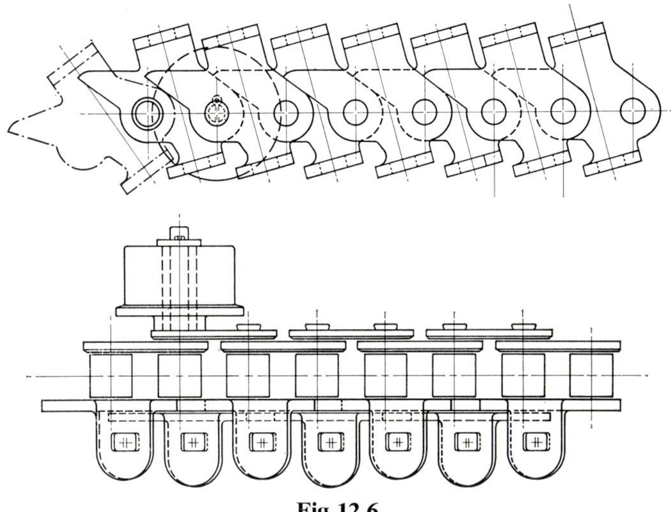

Fig 12.6

In the case of transmission chain, the roller diameter does not give any projection below the chain plates. It is thus desirable on applications of this type to fit support tracks between the chain plates, although chain running on plate edges is satisfactory if there is only light load on the chain.

Plate 93 shows the paper draw-off conveyor incorporated in a 2-colour offset litho printing machine using two transmission chains. As each printed sheet is about to be ejected from the printing rolls, the chains bring a gripper bar into close proximity to the rolls; a stationary cam track then opens the gripper jaws, which receive the sheet. The sheet is then drawn along by its leading edge, usually supported on tapes, until the stacking position is reached. At this point a second stationary cam opens the jaws, and releases the sheet. The operational speed may be as high as 300 ft/min.

The caterpillar cable drawing assembly shown in Plate 94 is used for the manufacture of electrical cable. The core of the cable is fed into an extruding machine, where it is coated with insulating material and then forced out through a die, which thus controls the external diameter of the insulation. The process is continuous, and constant tension is required as the cable is fed into the extruder. For this purpose, the caterpillar device grips and propels the cable by means of two parallel sets of rubber treads, each set being mounted on a pair of adapted transmission chains. The speed of the two sets of treads are synchronised and are also adjustable in height to accommodate various sizes of cable. Each rubber block is secured to the chains by pins passing through the block at the underside.

A simple conveyor shown used in the manufacture of fluorescent tubes is shown in Plate 95. This employs a 0·5 in pitch transmission chain having K1 attachments on both sides at every 4th pitch. Fixtures are bolted to these, on which the stem and exhaust tube elements are conveyed to the butt sealing process.

12.2 Special purpose chains

SLAT BAND CHAIN

This chain (see Plate 96), is popular for the dairy, brewing, food and similar industries in which a great many applications require the use of a steadily moving flat and level platform, on which bottles or similar containers can be conveyed in safety and under hygienic conditions. A range of slat widths from 2·5 in to 7·5 in is covered by BS 2075:1971. These are manufactured in mild steel or stainless material if resistance to corrosion is required. The maximum allowable pull for narrow chains with a single hinge is 550 lbf and 1,000 lbf for 7·5 in wide chain which has a double hinge. This pull must not be exceeded otherwise the gearing barrels will tend to open out, resulting in the bearing pins becoming loose and the barrels will no longer gear accurately with the wheel teeth. The chain speed is usually a maximum of about 150 ft/min, though some conveyors have been successfully run at greater speeds.

The wheels have teeth machine cut at half pitch spacing, and there are an odd number so that each tooth is in engagement with the chain once only in two revolutions, thereby doubling the life of the wheel.

The top run of the chain should be supported on smooth and level guides spaced 1·75 in apart for single hinge chains and 3·3 in for double hinge. With these dimensions the guides will not only support the chain, but will constrain it against side movement by contacting the sides of the gearing barrels as shown in Fig 12.7. The tops of the guides must have a lead-in to the wheels at their extremities, and the height dimension above the horizontal axis of the wheel must be adhered to, as shown in Fig 12.7. The return strand of the chain is preferably supported on continuous guides. If support rollers are preferred, they should have a diameter of not less than 7 in, otherwise there will be excessive continuous back bending which may tend to open out the gearing barrels. To minimise catenary pull the rollers should be spaced not less than 30 in apart.

When the application requires avoidance of contamination, a liquid soap solution can be used as a lubricant. An alternative is the employment of water soluble oil. Continuous lubrication can be effected by running the return strand through a bath, or by arranging a suitable type of drip feed.

Two conventional and widely used applications are illustrated in Plates 97 and 98. The former shows the feed end of a bottle washing machine having two parallel strands of chain for different circuits. Plate 98 shows a conveyor employing four parallel strands of chain carrying bottles from a de-crater to the washer. The unusual application shown in Plates 99 and 100 indicates that slat band chain usage is not confined within the industries already outlined. This installation employs the chain to move automotive engine pistons to and from a grinding machine. Part of the circuit is inclined and slipping of the pistons is prevented by studs welded to the slats. The slats are hardened mild steel, which have resistance to abrasive wear.

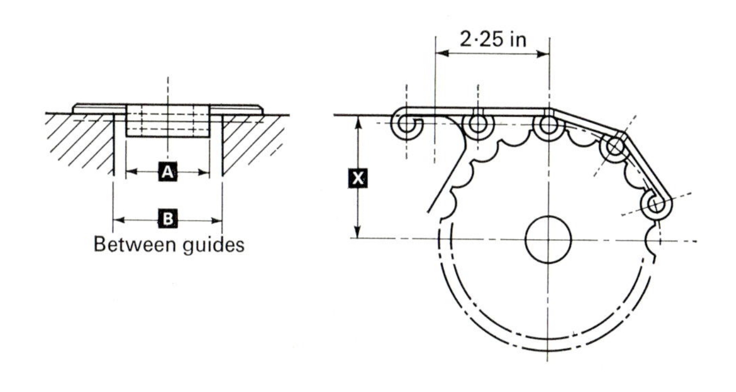

A Single hinge 1·665 in max.
Double hinge 3·160 in max.

B Single hinge 1·75 in max.
Double hinge 3·30 in max.

No. of teeth	19	21	23	25	27	29	31
X*	2·37	2·61	2·85	3·09	3·33	3·57	3·81

* *Limits of Tolerance for X measurement=0·02 in*

Fig 12.7 Installation details for slat band chain

BIPLANAR BOTTLE SLAT CHAIN

This is another special type of chain constituting a slatted platform, but biplanar joints permit articulation in two planes, on the principle shown in Fig. 12.8. The slats of stainless steel approximate to a crescent shape, and are mounted on 3,000 lbf breaking load conveyor chain with zinc plated components. The slats are riveted to the extended bearing pins of the outer links. Guiding round bends is effected by wheels of the required diameter, the tooth design of these wheels depending on which plane is articulating at that point. Wheels should have a minimum of 8 teeth and the chain must be supported throughout on guide tracks, this applying also to the return strands. Lubrication is achieved in the manner described for slat band chain. The use of this chain is illustrated in Plate 101 which shows a biplanar conveyor moving whisky bottles through filling and capping machines, to the labelling and packing benches.

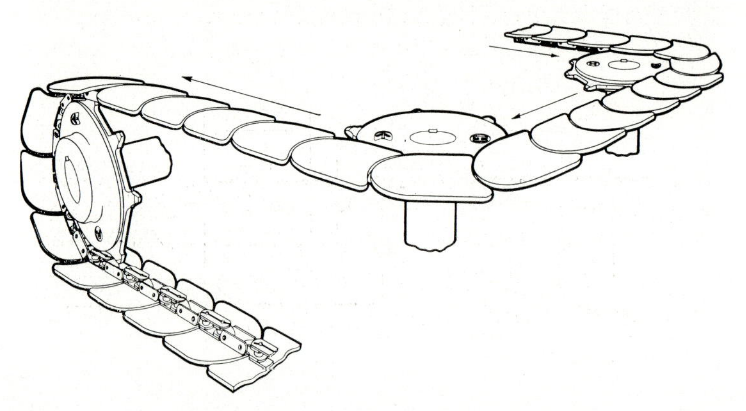

Wheels are marked H and V to signify the plane of operation, and must be fitted with the arrow pointing in the direction of rotation. When one-sided boss wheels are used, the position of the boss must be considered.

Fig 12.8 Biplanar bottle slat chain

AGRICULTURAL CHAIN (BS 2947:1970)

The available range of Agricultural chains used for conveying is detailed in Table 12.1 and illustrated in Fig 12.9.

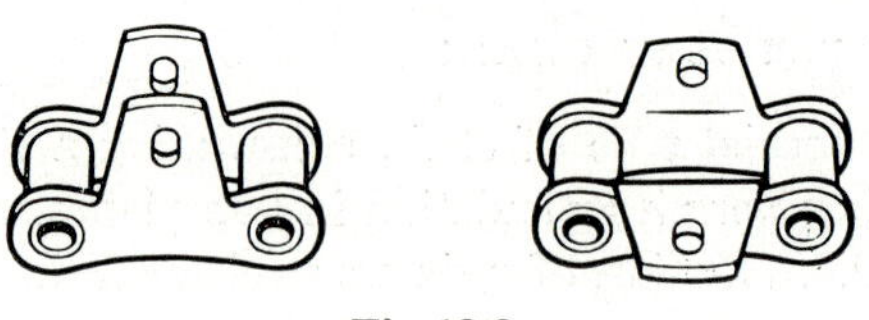

Fig 12.9

Plate 91

Plate 92

Plate 93

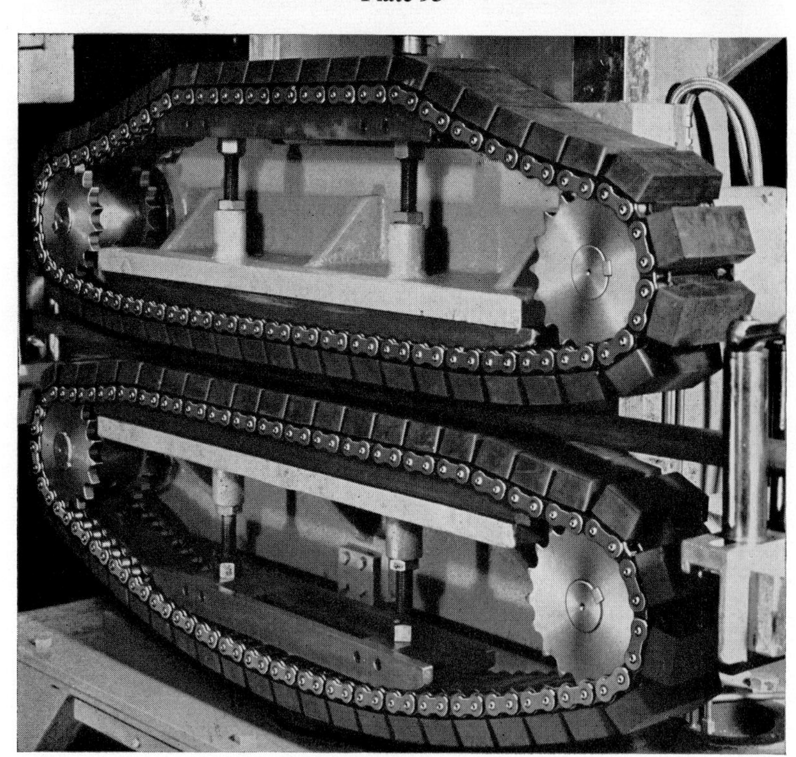

Plate 94

Plate 95

Plate 96

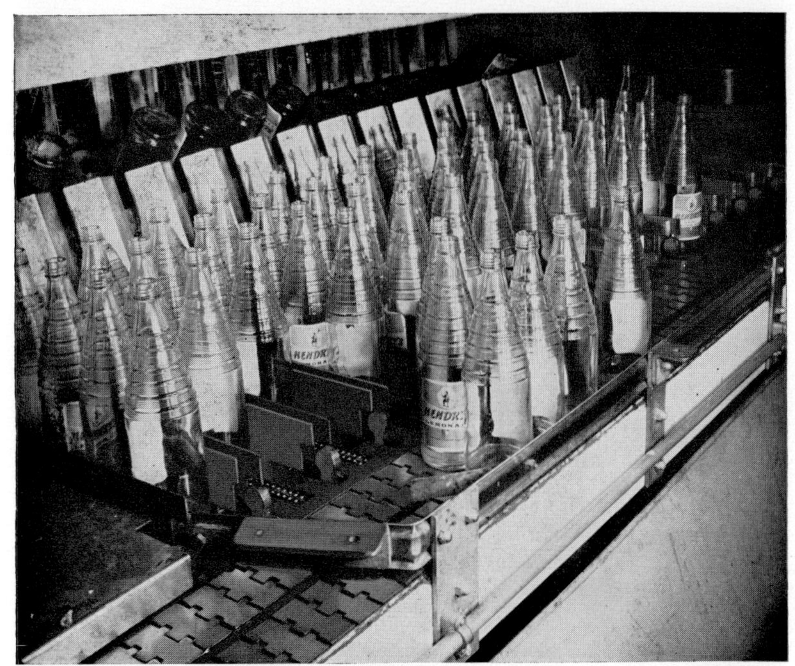

Plate 97

Plate 98

Plate 99

Plate 100

Plate 101

13

Chain Wheels

General description 13.1

The normal function of a chain wheel is not only to drive, or be driven by the chain, but wheels are also necessary both to guide and to support the chain in its intended path.

Wheels manufactured from good quality iron castings are suitable for the majority of applications. For arduous duty, it may be necessary to use cast steel wheels having a 0·45% carbon content. For extremely arduous duty the tooth flanks should be flame hardened. There are other materials which may be specified for particular requirements. Stainless steel for example is used in high temperature or corrosive conditions.

Depending on the size of wheel and loading, the design may be solid or spoked, the latter having side strengthening ribs incorporated where necessary. A typical one piece cast wheel is shown in Fig 13.1.

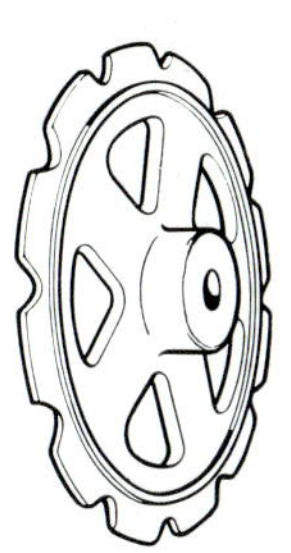

Fig 13.1

If quick detachability is necessary without dismantling shafts or bearings then wheels may be of split type. These are made in two half sections and the mating faces machined to allow accurate assembly with the shaft in place. If necessary, separate teeth, or segments with several teeth, can be employed, and attached by fitted bolts to a cast iron or steel body. This is a useful feature when conditions are such that tooth wear is abnormal.

Shafts, whether they are through shafts or of stub type, should be of such proportions and strength that wheel alignment remains unimpaired under load. Shaft sizes should be selected taking into account combined bending and torsional moments as shown in Appendix 2.

13.2 Wheel dimensions

Salient wheel dimensions are shown in Fig 13.2.

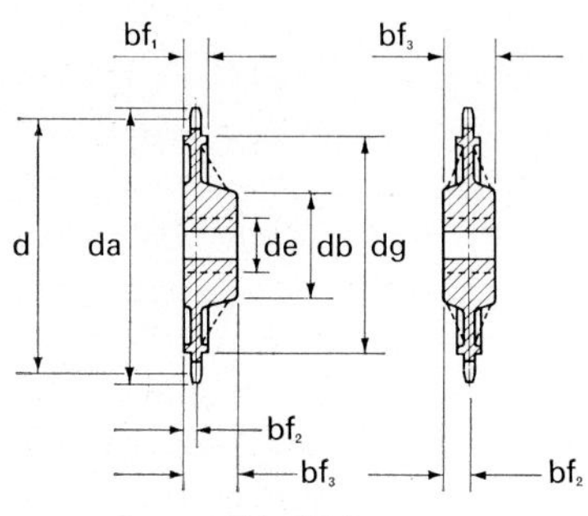

Fig 13.2

d $=$ Pitch circle diameter.
da $=$ Top diameter.
db $=$ Boss diameter.
de $=$ Bore diameter.
dg $=$ Shroud diameter.
bf_1 $=$ Shroud width.
bf_2 $=$ Face to wheel centreline.
bf_3 $=$ Distance through boss.

The pitch circle diameter, is a circle drawn through the bearing pin centres when a length of chain is wrapped round the teeth. Table 13.1 shows pitch circle diameters for wheels to suit a chain of unit pitch. The pitch circle diameters for wheels to suit a chain of any other pitch are directly proportional to the pitch of the chain.

13.3 Tooth form

For most applications the wheel teeth as cast and unmachined are satisfactory but machine cut teeth may however be preferable as referred to later. In conjunction with the chain rollers, the shape of the teeth facilitate a smooth gearing action. The tooth shape, whether cast or cut, is based on chain roller diameter and pitch for each specific chain. To ensure easy entry and exit of the chain the teeth are radiused at their outside faces, on the periphery.

In some handling equipment, such as elevators and scraper conveyors, both the chains and wheels have to operate in contact with bulk material which is liable to enter the spaces between chain rollers and wheel teeth, where the roller pressure can cause the material to pack between roller and teeth. If this is allowed to occur the chain then takes up a larger pitch circle diameter leading to excessive chain tension, and possibly breakage. This packing effect can be minimised by relieving the tooth gap as shown in Fig 13.3.

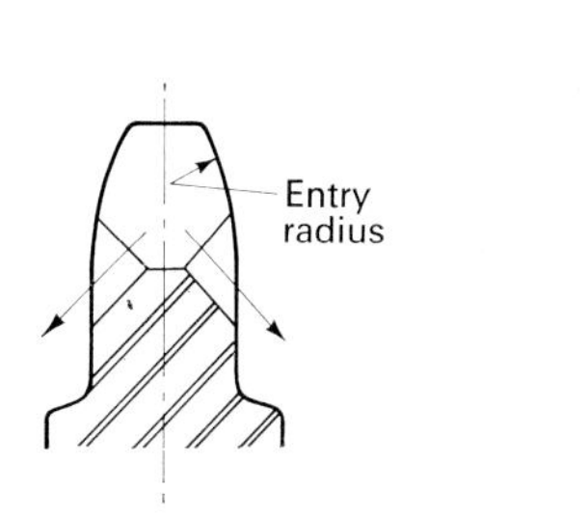

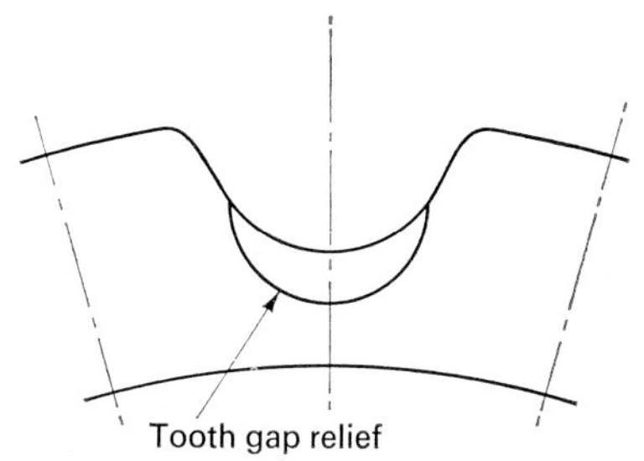

Fig 13.3

Machine cut teeth with their closer tolerances are employed in the class of applications listed below because of their greater accuracy.

High speed applications with chain speed in excess of about 180 ft/min.

Where synchronisation of the chain to a predetermined stopping position is required, with the angular wheel movement as the controlling mechanism.

Where numerous wheels are employed in a closed circuit, and variations in tooth form and pitch circle diameter could result in a tendency to tighten or slacken the chain on straight sections. This applies particularly where the wheels are closely spaced in either the horizontal or vertical planes or in close proximity in combined planes.

Where the linear chain speed variation has to be reduced to a minimum.

Number of teeth **13.4**

For the majority of conveyor applications experience shows that 8 teeth represents a reasonable minimum size for wheels. Below this the effect of polygonal speed variation is pronounced (see Section 13.5). Table 13.2 indicates the normal range of wheels for conveyors and elevators.

13.5 Chain speed variation

All chain wheels are in effect polygons having a number of sides equal to the number of teeth. When wheel rotation takes place, the chain, on engagement and disengagement, rises and falls relative to the wheel axis. This is shown in Fig 13.4.

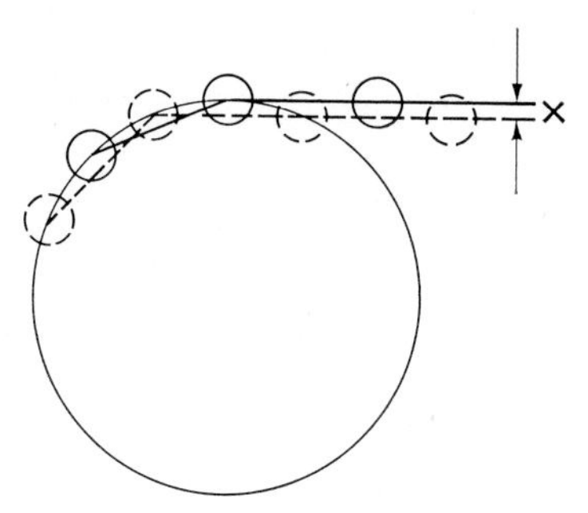

Fig 13.4

Actual variations are given in Table 13.3. These are based on unit pitch chain therefore 'x' should be multiplied by the actual chain pitch to obtain the finite dimension.

From the table it will be seen that the radius at position A is greater than that at B. Therefore if the wheel rotates at a uniform angular speed then the linear speed of the chain will be a maximum at A and minimum at B.

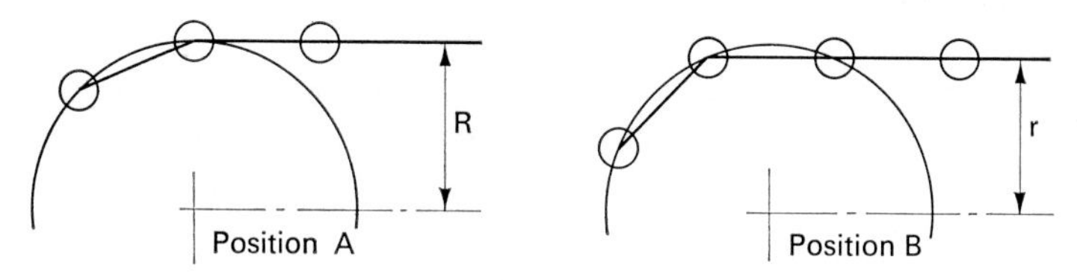

Fig 13.5

The percentage speed variations for typical wheels are shown in Table 13.4:—

For example a chain operating on an 8 tooth wheel at a nominal speed of 100 ft/min, will be subject to a theoretical speed variation between 92·4 and 100 ft/min.

Chain and wheel layout

In a conveyor or elevator where two chains and four wheels are employed, it is normal practice to keyway the driving wheels to the shaft as a pair, with the teeth in line to ensure equal load sharing. When a pair of wheels is mounted on a shaft the long bosses of the wheels should be assembled so as to face each other, i.e. to the shaft mid point. This allows the wheels to lie close to the shaft bearings, giving maximum support to the load, while at the same time requiring only the minimum width of the conveyor structure. It is usual to secure the wheels to the shaft with gib-head keys, the key head being at the long boss end. Non-driving or tailwheels on the same shaft are arranged so that one wheel only is keyed to the shaft, the other being free to rotate, so accommodating minor differences in phasing between the chain strands. This wheel should be located between fixed collars secured to the shaft on each side. The free tailwheel may have a phosphor bronze or similar bush, but generally this is not necessary since the relative movement between the floating wheel and shaft is small.

In the case of more complex installations, such as two chain conveyors having several stages, the floating wheels should alternate from one side to the other along the circuit path. In this way the slight increase in load pull imposed by the effort of turning the shaft is distributed more evenly over both the chains.

14

Corrosive and High Temperature Applications

14.1 Corrosive applications

Every application involving a hostile environment demands a clear appraisal of all factors so that the installation will work with reasonable durability with the most economic design. Such environments may range from those allowing the use of normal chain to those where conditions approach the limit of feasibility and often demand a sophisticated approach in chain selection.

Where chains are in contact with pure water (pH value 7), town water or alkaline solutions (pH 8 to 14) at ambient temperatures and product contamination is not important then standard chains in normal materials are suitable. These have the advantage of low cost and ready availability. Where abrasion is present the harder surfaces obtained with case-hardened components, compared with the softer stainless materials, give good abrasive resistance. To give reasonable life a higher strength of chain is often used compensating for the effects of corrosion, this being a more acceptable proposition compared with a chain made from stainless materials. To reduce the effects of corrosion, standard chains may be zinc or cadmium plated, but since these are toxic their contact with food should be prevented.

Alkaline solutions produce a cleaning action which removes the lubricant from the chain. On these applications it is advisable to lubricate the chain with a de-watering oil.

Where chains are in contact with acid solutions (pH 1 to 6) then it is generally necessary to use chain with components manufactured from corrosion resistant materials. For the majority of applications a stainless steel material is used, the choice of which being the responsibility of the chain manufacturer.

Wheels manufactured from cast iron will give an acceptable life. For very corrosive applications then corrosion resistant materials would be specified but cost is higher and delivery extended.

To enable the chain manufacturer to select suitable materials, the following information is required:—

Description and arrangement drawing of the conveyor system,

with load data and particulars of proposed operation.

Details of all the substances present, including any corrosion inhibitors; noting that general statements such as 'acid', 'alkaline' etc are not sufficient. The chemical composition including pH value should be stated.

Operational and shut down temperature.

Whether contamination of the conveyed product can be tolerated or not.

The extent to which the solution or material handled will contact the chain or wheels, e.g. total immersion, partial immersion, in and out of tanks, splash or vapour etc.

Whether material handled, if solid, is dry or wet.

If the material is normally dry, is condensation prevented at shut down.

If the chain is a replacement, description of previous chain and details of service given.

Details of base chain proposed, along with type of attachments and preferred method of fixing.

14.2 High temperature applications

Chains made in normal materials can operate at temperatures up to 300°C but some softening of case-hardened surfaces will occur, this commencing about 200°C. To offset this effect the selection factors given in Chapter 2 should be increased by 25%.

For operation at temperatures from 300°C to 450°C stainless steel chain is used. For temperatures from 450°C to 800°C then heat resisting steel must be used. As the force fit between chain plates, pins and bushes is destroyed when chains are subjected to temperatures over 300°C it is necessary to 'flat' the pin and bush to prevent rotation, see Fig 14.1. Prevention of side movement of the link plates is achieved by riveting over the ends of the pins and bushes.

For temperatures above 850°C it is necessary to use a 2×1 block type chain manufactured from heat resisting steel. This type of chain is shown in Fig 14.2.

On all applications over 300°C full applicational details should be given to the chain manufacturer in order that the correct design of chain and suitable materials are used.

The information required is as follows:—

The maximum temperature to which the chain will be subjected. If this cannot be given then the anticipated oven temperature should be stated.

A description and arrangement drawing of the proposed installation, with data including chain speed, loading and type of chain attachments.

Whether there is actual flame impingement on the chain.

If the temperature change is intermittent, does cooling take place rapidly, as in quenching.

Nature of surrounding gases, and whether oxidising or corrosive (weakening by reduction) in character.

Method of heat application, and in the case of gas heating the fuel used, along with an indication of sulphur content.

Nature and concentration of any other substance present.

Wheel position in relation to the source of high temperature.

If the chain is a replacement, description required of previous chain, along with its operational record and length of service.

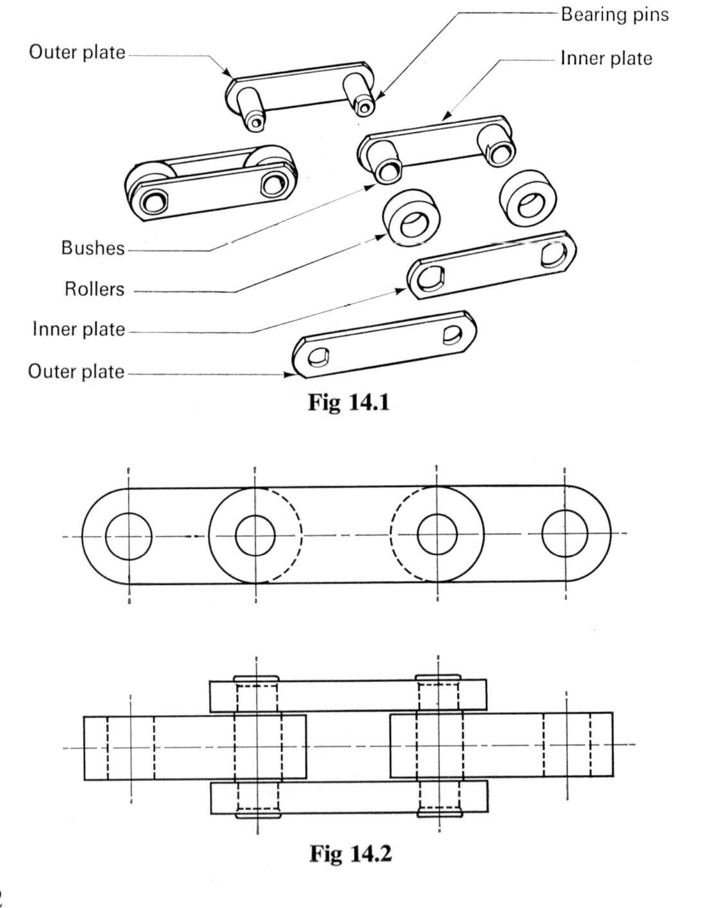

Fig 14.1

Fig 14.2

Nuclear radiation 14.3

Conveyor chains are installed where nuclear radiation is involved. Such applications are special and careful selection of suitable materials is called for depending upon radiation/temperature/corrosive environment. Full applicational details must be given to the chain manufacturer in order that all possibilities can be considered in the compilation of their design proposals.

15

Conveyor Drives

15.1 General

It is normal to drive a conveyor at the position of maximum load. In a simple horizontal or inclined slat conveyor this position coincides with the delivery or off-loading end. If driven at the tail or feed-on end of the conveyor, the chain pull would be marginally greater.

15.2 Driving methods

TOOTHED WHEELS

This is the normal method of driving a conveyor chain. Ideally the chain lap around the wheel should be 180°, if this is not possible then the aim should be to have a minimum of 3 teeth in gearing engagement. Where more than one chain is involved, it is usual for the driving wheels

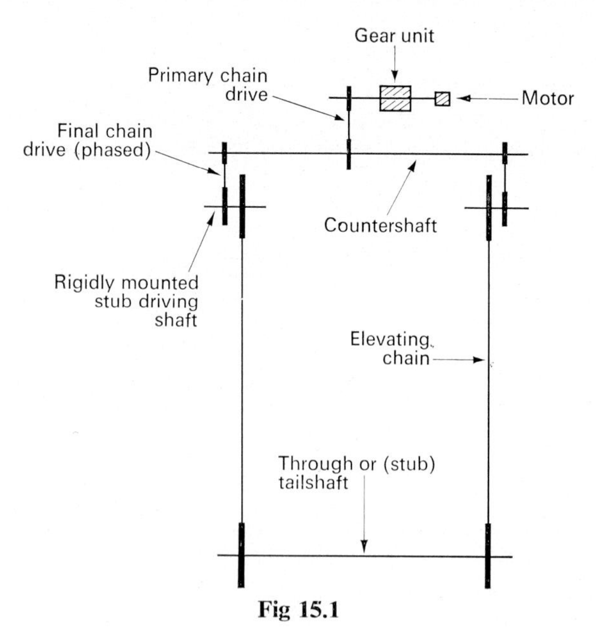

Fig 15.1

to be keyed onto a common shaft, so that there is no relative movement between the wheels.

On some types of swing tray elevators, a through headshaft is not possible because of the need for free passage of the trays and load between the wheels. In such cases, each wheel of a pair would be stub shaft mounted, the drives to the pair being accurately synchronised by the use of a counter shaft having a central drive from the motor, and chain drives from each end to the headwheels, as shown in Fig 15.1.

In this arrangement, torsional deflection of the countershaft in theory has no effect on synchronisation, due to the central input drive. However, it is possible to ensure synchronisation over long periods by introducing a phasing jockey system on each of the final chain drives to the headwheels. Adjustment of the drive chains is also an inherent feature of this layout.

Tailshafts may be of stub type for tray clearance as mentioned, or if a through tailshaft is used, the normal method of keying one of the conveyor chain wheels to the shaft allowing the other to remain free to follow slight chain phasing differences.

TOOTHED WHEEL TANGENTIAL DRIVES

When a standard wheel tooth form is used, a single wheel engaging the chain tangentially is not usually satisfactory even when the teeth are accurately cut to reduce to a minimum the free movement with the chain. As one tooth leaves its engagement there is an inherent angular distance lag before the next tooth gears into the chain and takes the driving load. The resultant backlash introduces uneven operation and there is always the liability of mal-gearing. It is possible to have two wheels in line at each drive point, these being interconnected by means of a differential or balancing gear to ensure load equalisation. On some overhead conveyors a single driving wheel is geared into the chain at a suitable point on one of the straight sections, adequate tooth engagement being achieved by humping the chain over a limited arc of the wheel periphery.

A unique example of a tangential drive in association with normal toothed wheel drive is now described. The tangential drive, located in the circuit to equalise the chain pull at each drive position, constitutes a booster drive. Plate 102 shows the system under test, its mode of operation being comparable with that of the caterpillar drive described later. The tangential engagement wheels were cut with a special tooth form in order to ensure gearing action. Tests established that because of the variation in direction of tooth loading pressure during the angular wheel movement, there was a tendency for each tooth to pluck the chain from the straight path, in addition to the expected outward reaction force. As this plucking force is of low value, constraint of the chain by means of guides above and below solved the problem. The guides form a plain continuously lubricated track, the construction of the backing guides are of interest as these comprise standard transmission chains fitted with K attachments to which are fixed recessed steel pads. The latter engage the plate edges of the main conveyor chains. Sliding friction is thus virtually eliminated and unimpeded gearing of the tangential wheels with the chain rollers is established.

CATERPILLAR DRIVES

The caterpillar drive can be defined as a toothed wheel of infinite radius which drives the conveyor chain by tooth engagement on the same principle as a tangential driving wheel. As shown in Figs 15.2 and 15.3 the caterpillar drive comprises a short centre drive unit having two equal sized wheels, the chain being fitted with driving dogs or teeth gearing with the main conveyor chain. This engagement takes place over a short length of the main chain where it runs parallel to the caterpillar chain.

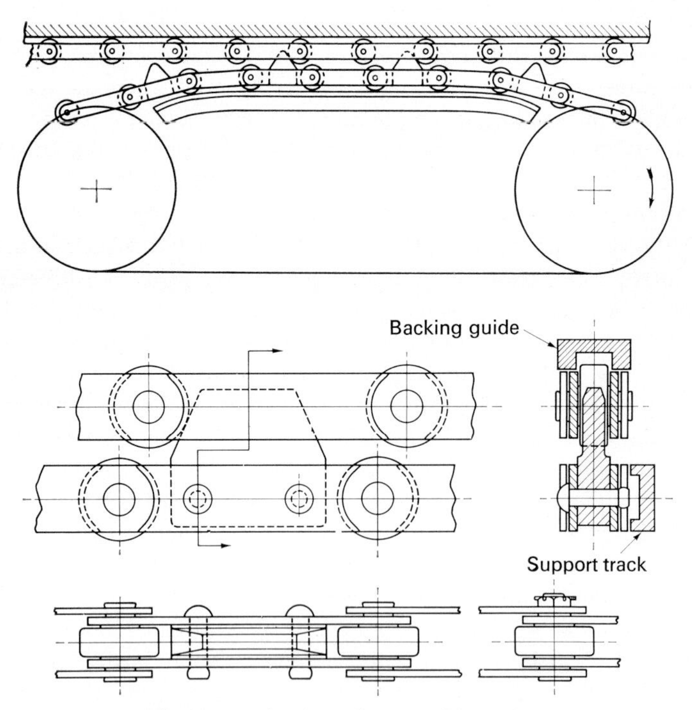

Fig 15.2 Fixed tooth caterpillar drive

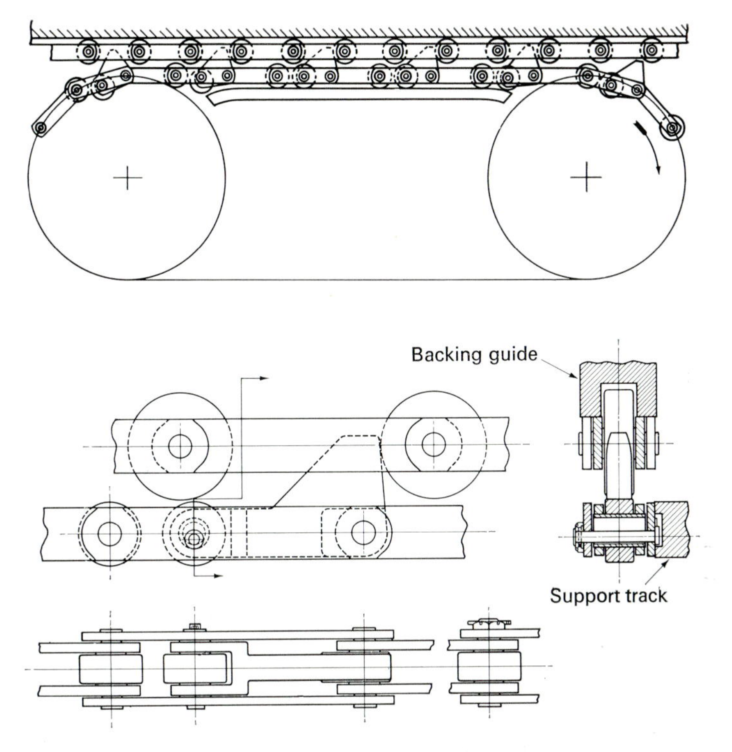

Fig 15.3 Hinged dog caterpillar drive

There are two types of caterpillar drive, these being respectively the fixed tooth (Fig 15.2) and hinged tooth (Fig 15.3); the former is most commonly used. The hinged tooth has the advantage that the gearing reaction pressure is taken by a roller fitted at its trailing end. Applications of these two types of caterpillar chain are illustrated in Plates 44 and 49, in both cases the chains run in the horizontal plane.

The caterpillar chain must be supported both on the driving and return strands from below, in order to avoid catenary sag which would prevent the caterpillar teeth from engaging the conveyor chain correctly. The latter must also be supported on recessed guide tracks, which must extend along the whole of the caterpillar drive engagement section, plus at least four chain pitches before and after this, to ensure that the gearing action is fully controlled. Figs 15.2 and 15.3 clarify this requirement.

The tooth driving faces must be tapered so that the reaction load is then borne equally at each pitch engagement by both the caterpillar chain and the main conveyor chain. The conveyor chain backing guides should bear on the chain plate edges and must not contact the rollers. These guides should be well lubricated at all times, and provision for ready replacement due to wear is necessary. In place of guides a series of closely spaced rollers of small diameter can be used, these again running on the conveyor chain plate edges.

Where the main conveyor chain is of large pitch, a caterpillar drive could be used comprising a smaller pitch chain running on wheels having a relatively large number of teeth. This reduces the polygonal effect, since to obtain equivalent smoothness of the main chain a large toothed wheel would be required which might be difficult to accommodate. This smoothness of operation is of importance in many industries, for example in continuous casting in moulds carried by the conveyor system.

INTERMITTENT DRIVES

On some conveyor systems, intermittent rather than continuous motion is required. A good example is shown in Plate 97 where on a bottle washing machine the bottles are dropped into pocketed slats carried by the chains, the latter at that moment being stationary.

Another widely used application is found on apron feeder conveyors described in Chapter 5, when these are employed to feed crushers by a controlled interrupted feed motion.

Intermittent motion can be obtained from the continuously rotating shaft of the prime mover, by the ratchet and pawl principle; this has been used for many years. The preferred modern equivalent, providing much higher standard of accuracy of movement as well as total enclosure, is the Sprag Clutch which is widely used as an indexing medium.

Hydraulic movement of the conveyor chain is also employed. In this case, the motion being imparted to the chain by means of a driving dog, which is mounted on a hydraulic ram.

15.3 Multiple drive points

The general purpose of a multiple drive system is to reduce the load pull in the conveyor chain, enabling a smaller size of chain to be used.

SINGLE PRIME MOVERS

On many conveyor circuits, such as a bread cooler (see Plate 71), circuit layouts take up little space but afford maximum travel. As a result the chain wheels are close together, enabling a single prime mover to be used to drive a number of shafts. In the two examples shown in Figs 15.4 and 15.5, the driving wheels (DR) can be connected together by transmission chain and gearing; the former being normally used.

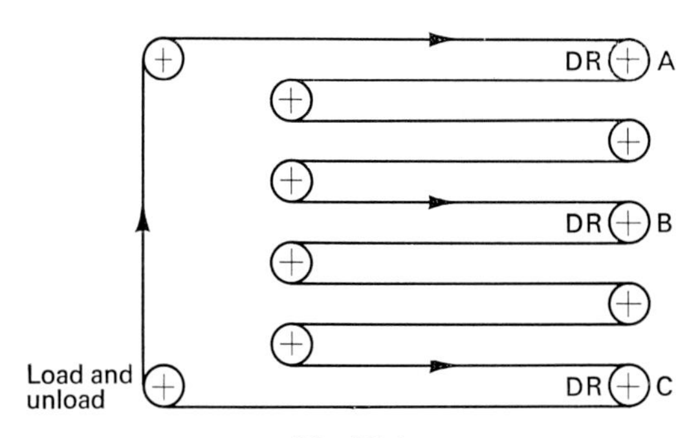

Fig 15.4

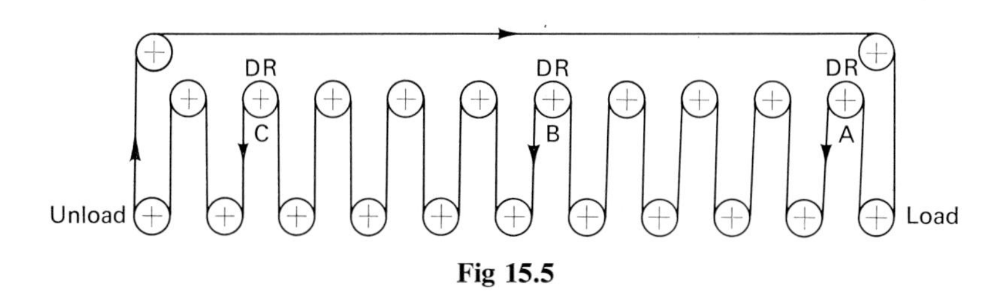

Fig 15.5

Fig 15.6 shows four separate chain drives, with the driving shaft at the centre. With this arrangement the inner chain drives will transmit double the power of the outer drives, assuming equal loading on all four driven shafts.

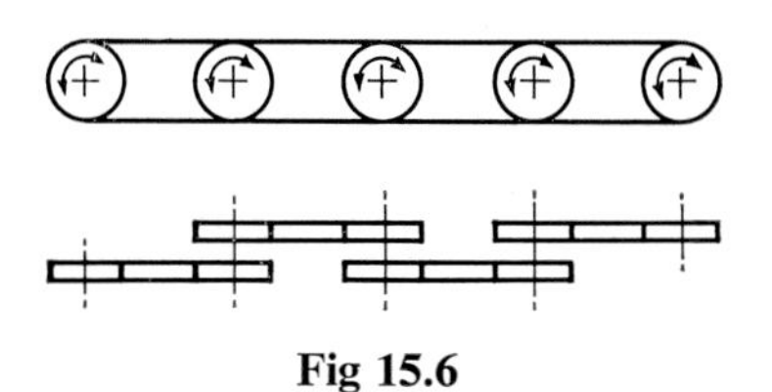

Fig 15.6

Fig 15.7 shows an arrangement having a single chain, which is made to gear with the series of driven wheels (DR) by use of a number of jockey wheels (J). Correct chain tension is maintained by an adjustable jockey.

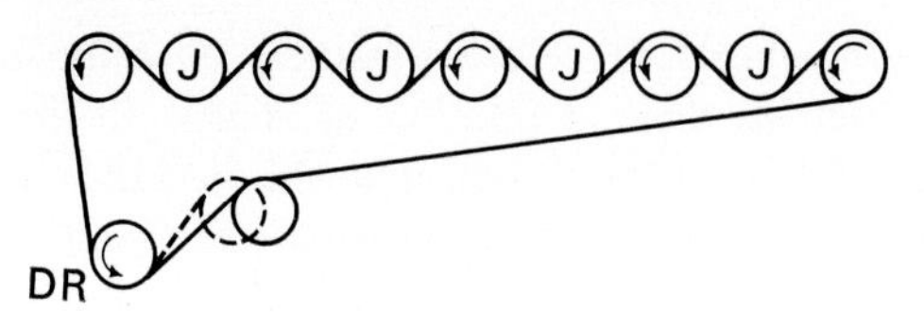

Fig 15.7

If the chain drive spans a considerable distance between shaft centres, (e.g. 10 ft or more) both strands of the transmission chain should be supported as shown in Fig 15.8. It will be noted that the raised guide track supports the chain rollers. The method is applicable providing the lie of the drive is at only a small angle to the horizontal, and its linear chain speed does not exceed 200 ft/min.

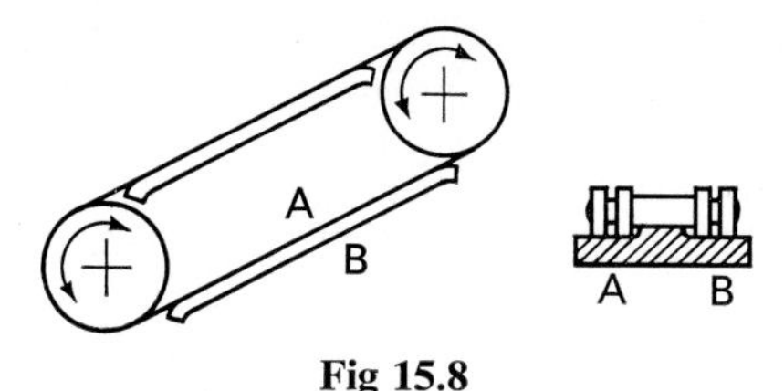

Fig 15.8

MULTIPLE PRIME MOVERS

On conveyors using more than one prime mover cage motors having a fairly high slip characteristic can be used. The motors are connected to a common starter.

On complicated conveyor circuits the torque variation may be considerable. In such cases, it is usual to employ a slip coupling in each drive such as fluid, particle, or electro-magnetic types.

15.4 Drive arrangements

The drive should be a well engineered sequence transmitting power with optimum efficiency at the most economical price. Any drive must suit requirements of space, disposition of shafts, cost and delivery whilst utilising wherever possible stock transmission components.

The main components of transmission drives, listed in order of drive sequence, are:—

PRIME MOVERS

For the majority of industrial applications a totally enclosed fan cooled cage type AC electric motor is used. To minimise motor cost and obtain maximum drive efficiency a 4 pole electric motor is usually preferred.

Electric motors as described above are not intended for frequent starting. If the driven machine has to be started more than say 20 times per hour then some form of clutch may be necessary.

V-BELT DRIVES

These are used principally as a means of shaft connection and speed reduction between an electric motor and a gear unit or variable speed unit. With a V-belt drive there will be some belt 'slip' but this can usually be ignored unless a precise and positive ratio must be maintained. A stock V-belt drive together with a stock gear unit can usually be selected to give the overall ratio desired.

INFINITELY VARIABLE SPEED DRIVE UNITS

There are three basic types of variable speed units, i.e. Mechanical, Hydraulic and Electrical.

However, many forms of these basic types are available resulting in considerable difficulty in narrowing down the choice. To assist in drive selection, characteristics of the main types are given.

Mechanical belt drives

These consist of two or more pulleys connected by belts. The belt may be either standard or extra wide V-section or slatted depending on which drive is selected. Speed variation is obtained by adjusting the diameter of one or more of the pulleys. Belt drives are widely used, being a reliable and simple method of achieving a variable speed drive, and do not require skilled maintenance.

They are somewhat restricted in their speed ratio and in no case is it possible to regulate the speed down to zero. Since belt slip is unavoidable (except in the PIV unit) precise speed regulation is not possible, although in many applications this is not very important. It is recommended that the drive be run throughout its full speed range once every day so that distribution of lubricant prevents fretting damage between pulleys and shaft.

Normally a belt type with the lowest possible speed range should be selected. To select a drive with a higher speed range than necessary will probably involve greater capital cost and give reduced efficiency and life.

Hydraulic

Drives using oil or similar materials as the power transmitting medium can be sub-divided into distinct types which have quite different characteristics. i.e. Hydrostatic and Hydrokinetic, but the latter type is little used for variable speed conveyor drives although the traction type hydraulic coupling is often used because of its full load start/slip facility.

The hydrostatic drive comprises a positive displacement pump driven by a power source and supplying accurately controlled rates of oil flow to a positive displacement motor from which the output drive is taken. Power transmitted is proportional to the rate of oil flow, which is determined by the input pump unit and to the effective working oil pressure, which is entirely dependent upon the resisting torque generated by the driven machine. Pumps, motors of fixed or adjustable displacement are generally of multiple piston construction. A typical unit is the Carter Hydrostatic Infinitely Variable Speed Drive (Fig. 15.9).

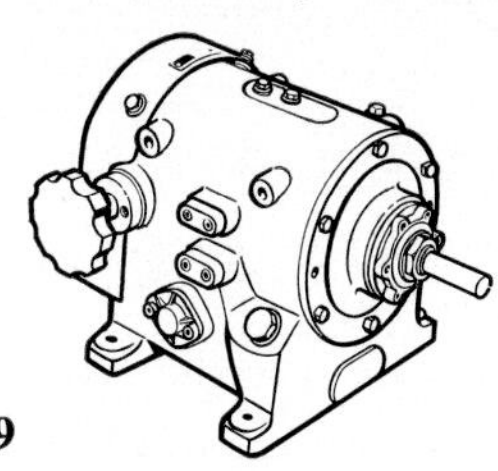

Fig 15.9

The main advantages of most hydrostatic variable speed drives are:—

Speed range around 27/1 ratio which is usually adequate for the majority of requirements.

Electrical simplicity. Hydrostatic drives normally use cage type electric motors which may have protected, drip proof, total or flame proof enclosures as required.

In-built dynamic braking characteristic without any electrical complexity. Whilst the input shaft is being driven, movement of the speed control provides acceleration or braking torque as required.

Output speed can be changed with the drive running or stationary, allowing controlled acceleration of the conveyor from zero speed even when the drive has been stopped at a high speed setting.

Ease of integration into process control schemes using standard forms of industrial servo equipment.

Continuous running for prolonged periods at any selected speed setting is possible. Hydrostatic drives do not normally suffer from tracking or fretting conditions which could affect their control.

Electrical

The development of solid state electronics has brought the Thyristor or SCR (silicon controlled rectifier) (Fig 15.10) into great prominence as the most versatile, reliable, compact and economic system for electrical variable speed drives particularly up to the medium horsepower range. The AC current is rectified to DC current by Thyristors, i.e. rectifiers such as silicon or selenium types. The DC current is then used for speed control of the DC drive motor. The electronic rectification unit can provide controls to maintain motor speed within 2%, 3% or 10% depending on the sophistication of the circuit against the varying load on the motor output, to give inching, to give dynamic braking, and provide protection against overload etc.

Units are available from fractional up to 400 hp and with maximum speed range of 100:1. The popular range is up to 7·5 hp and ratio 25:1.

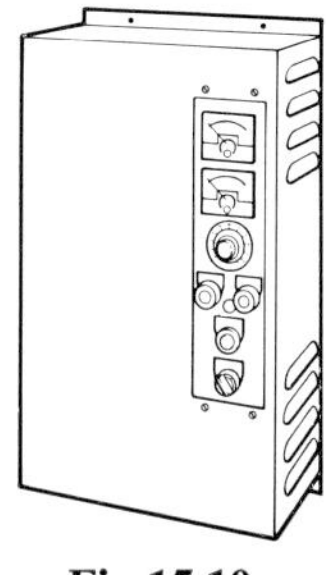

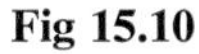

Fig 15.10

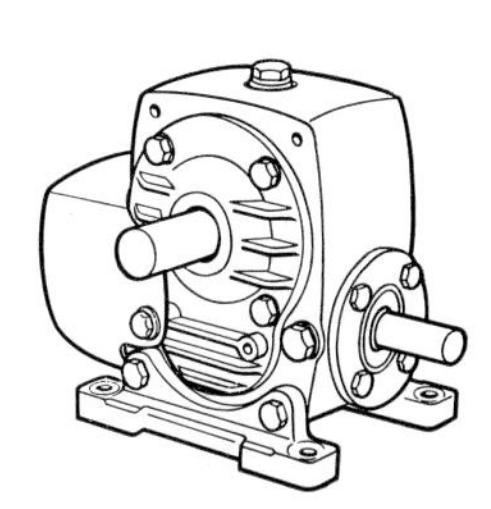

Fig 15.11

GEAR UNITS

A wide range of gear units is available but some of these will be eliminated when disposition of shafts, speed ratio and power output is taken into consideration.

Maximum efficiency is obtained from helical or spur gear units. Worm gear units (Fig 15.11) may have a lower efficiency at high ratios but are quieter and for high ratios are more compact than helical or spur units.

A geared motor unit, i.e. a gear unit with integral flange mounting motor, should be used where possible. This reduces the number of transmission items, lowers cost and simplifies drive installation. The range of geared motor units is limited, therefore it may be necessary to use a separate motor and gear unit. Where output speed is low and torque is high then many arrangements can be used. Notes on these, in order of preference, are given below:—

A double worm gear unit is ideal.

A spur/helical geared motor unit, as the primary drive, coupled to a worm gear unit. The reason for this arrangement being: stock or standard ranges of spur and helical gear units have smaller speed/power range than worm gear units. Spur/helical geared motor units are more readily available.

Within certain limits of ratio, a combined helical and worm gear unit with its higher efficiency but higher cost.

Some degree of flexibility is desirable in most transmission drives and therefore chain drives are not normally used on both input and output of gear units at the same time. Flexible couplings or V-belt drives are preferred for the input with a chain drive from the output shaft if required.

CHAIN DRIVES

As already mentioned chain drives are used from the output shaft of a gear unit to the conveyor headshaft, generally giving further speed reduction. Transmission chains present an efficient method of power at short and medium centre distances. The drive is positive and so preserves a definite speed ratio.

Roller chains are manufactured to ISO recommendation R.606 which incorporates BS and AS chains. Stock drives using BS chain and wheels are available to transmit up to 700 horsepower at 550 rpm and with ratios up to 7:1.

15.5 Safety and overload devices

On conveyors which start up under full load it is desirable to incorporate a fluid coupling or centrifugal clutch etc. This allows smooth acceleration of the system up to full running speed thus eliminating shock loading in the drive. The coupling is normally fitted between the motor and a gear unit.

Failure of a conveyor drive on an inclined conveyor or elevator can lead to runback of the load. This may occur due to the motor current being tripped because of overload, breakdown of current supply or failure of a drive component. To prevent runback occurring then some form of backstop should be fitted.

A legacy of the past but which is no longer valid was to rely on the inefficiency of the gear unit to prevent runback. Nowadays runback is possible even with high reduction worm gear units because of their high efficiency and therefore a separate safety feature must be employed. British standard 721:1963 Worm Gearing recommends the use of a separate backstop where runback can occur. Ratchets, electro-magnetic or automatic brakes are often employed for this purpose. With a ratchet or brake, some runback may occur before the device holds, and even a small amount may give rise to shock loading in the system when motion is arrested. A device which is now widely used is the sprag clutch which provides a superior solution to this problem. A typical sprag clutch unit is shown in Fig 15.12.

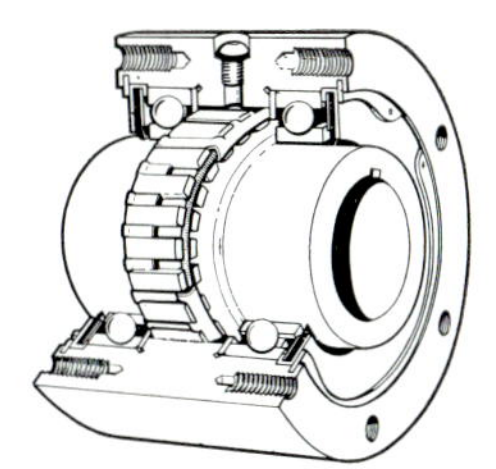

Fig 15.12

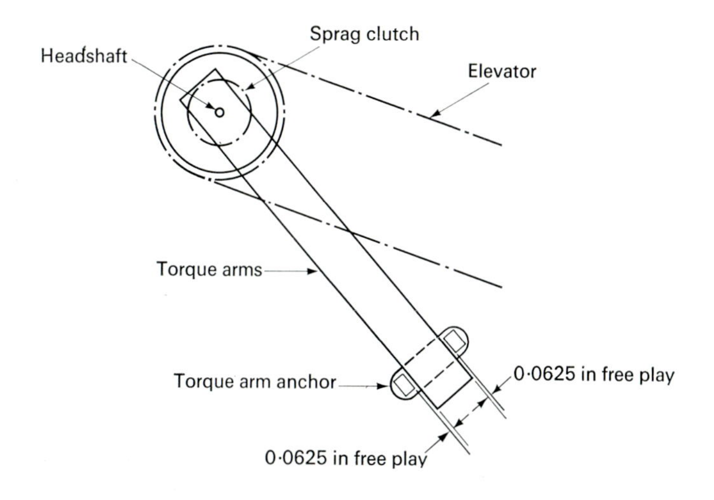

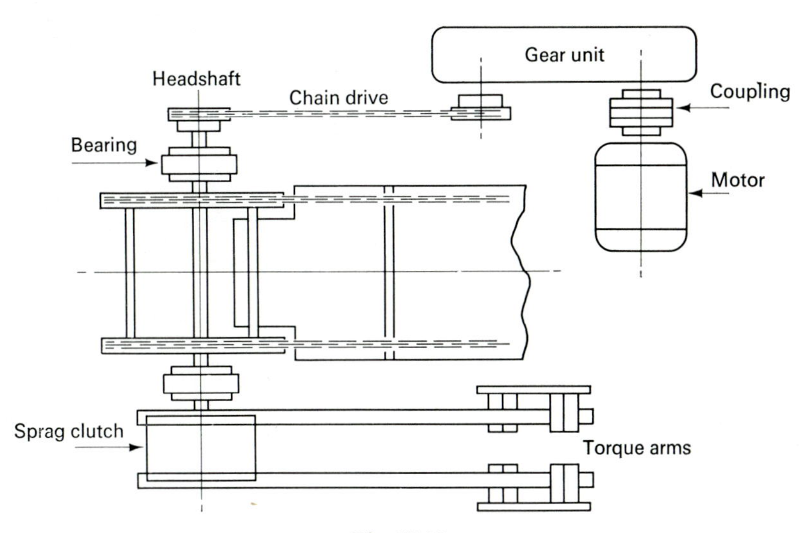

Fig 15.13

Fig 15.13 shows how a sprag clutch is employed as a backstop on an elevator. Here the inner race is keyed to the conveyor headshaft whilst the outer race is held stationary by torque arms which locate on the conveyor structure. The torque arms should not be rigidly fixed as it is important for the outer race to be free from axial twisting. Under normal running the inner race overruns but there is instantaneous hold-back with no measurable backlash if drive failure occurs.

Should breakage of the conveyor chain occur then it is possible to restrict the extent of chain travel by fitting hinged pawls on the loaded upward run of the conveyor. Normally the chains by-pass the pawls but when direction is reversed—as in the case of runback—then the pawls engage with the chain rollers and halt travel. Overhead conveyors can have pawl arrangements on inclined sections but these are made to contact the trolley rollers as shown in Fig 15.14.

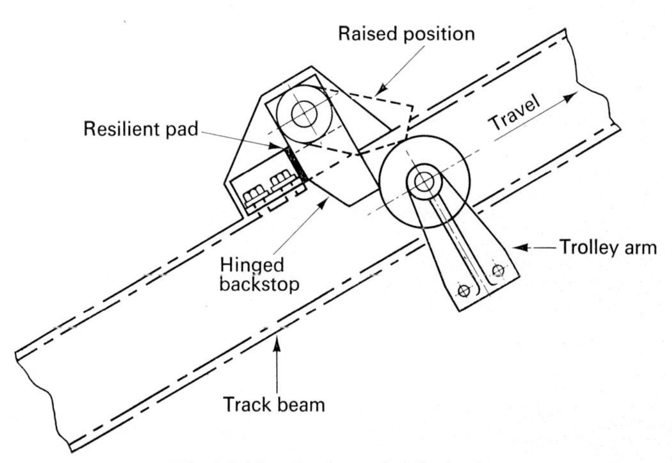

Fig 15.14 Anti-run-back device

On conveyors liable to shock load or overload then the drive should be protected by an overload device. This should preferably be located on the conveyor headshaft.

The best mechanical system intended for short duration overloads is the Torque Limiter. This is a compact overload device which slips at a pre-determined load. It automatically re-engages when the overload has ceased or the obstruction has been removed, no re-setting is required. Primarily it is intended for use with a chain drive but can also be used with gears and other transmission media. As illustrated in Fig 15.15 it comprises a fixed hub, loose bronze bearing, two friction discs, two pressure plates, conical spring washer/s and a ring nut. The chain wheel or other equipment is carried on the bronze bearing between the friction discs and the pressure plate. Electrical cut out can be incorporated.

Gripping pressure is achieved by compressing the conical spring washers.

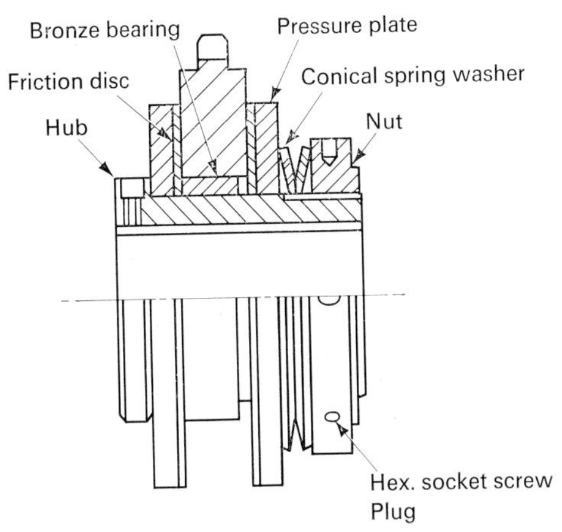

Fig 15.15

An alternative is the fluid coupling between motor and gear unit. Basically it comprises two rotating elements, an impeller and runner. Power is transmitted from the impeller to the runner from the kinetic energy built up in the oil rotating in the torus formed by the impeller and the runner due to 'slip' between the two. Such kinetic energy being transformed into mechanical energy, hence torque, when the oil transfers from the blades of the impeller to the blades of the runner. A typical fluid coupling is illustrated in Fig 15.16.

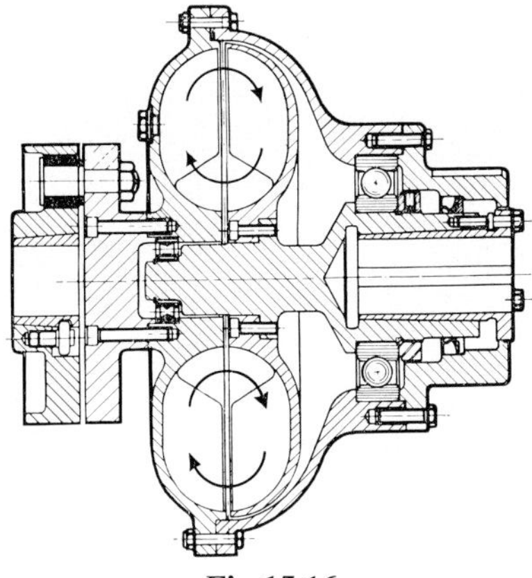

Fig 15.16

An alternative mechanical device is the shear pin. The adaptor is keyed to the shaft and the torque transmitted via the shear pin. Pins are made either from brass or steel and grooved to ensure controlled fracture and easy withdrawal after shear, they must cater for full starting load. Fig 15.17 shows a typical chain wheel incorporating a shear pin.

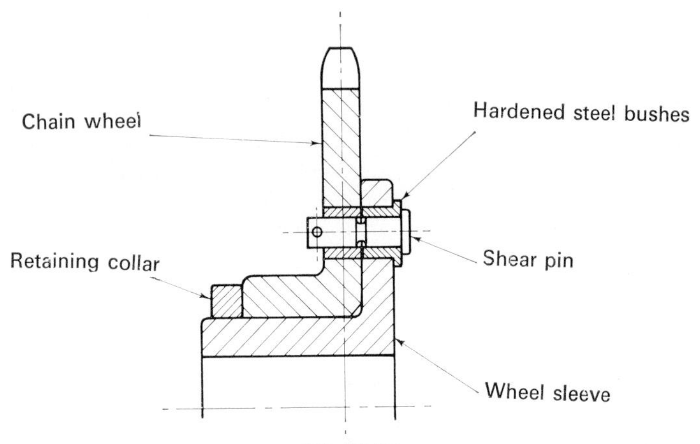

Fig 15.17

Plate 102

16

Chain Installation and Maintenance

Installation **16.1**

Dismantling, reassembly and repair of conveyor chains are greatly assisted by the use of the special tools available. The various items of equipment available are illustrated together with operational description. Fig 16.1 shows a ratchet type of cramp for bringing the ends of the chain together to allow insertion of a connecting link.

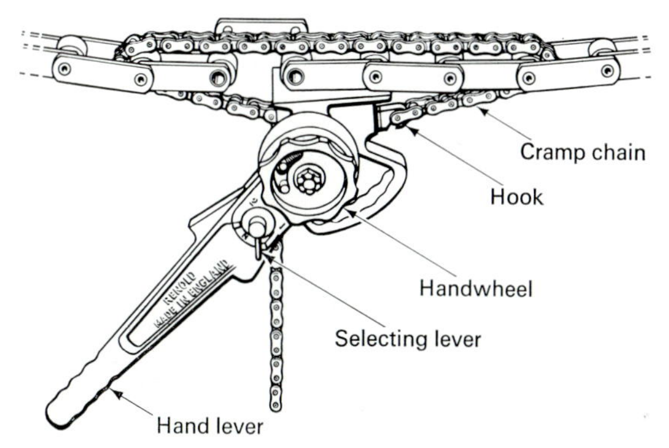

Fig 16.1 Chain cramp

Figs 16.2 and 16.3 illustrate two sizes of assembly tool suitable for solid and hollow bearing pin chains. Their function is to connect two inner links with an outer link or connecting link by forcing the loose plate on to the ends of the connecting pin.

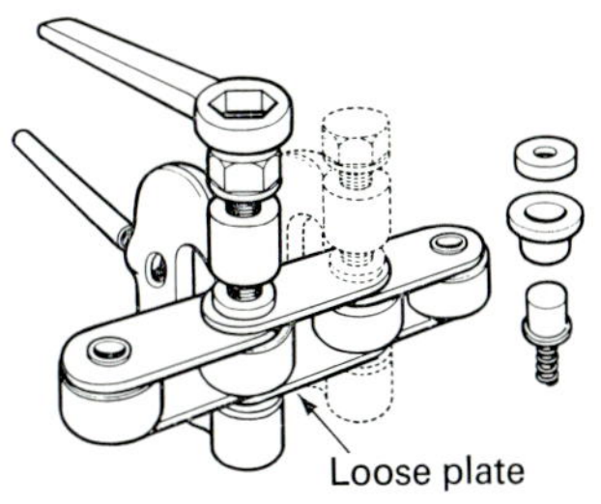

Fig 16.2 Assembly tool for use with hollow and solid bearing pin chain

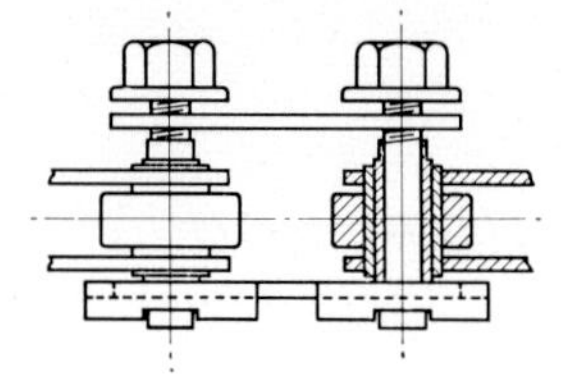

Fig 16.3 Assembly tool for use with hollow pin chain

Fig 16.4 illustrates an extractor tool, this being used to uncouple the chain at a riveted link by forcing out the bearing or connecting pins from the chain plate.

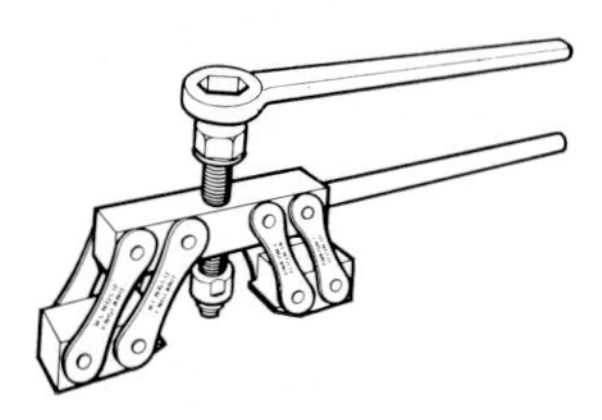

Fig 16.4 Extractor tool for use with hollow and solid bearing pin chain

Where conveyors are out of action for long periods, due to seasonal fluctuations etc., the chain should be removed from the conveyor thoroughly cleaned and coated with a protective lubricant. Spare chain kept in store should be similarly stored. If left exposed to the atmosphere without protection, chains of normal material may become corroded, particularly if there is condensation present. If it is not convenient to remove the chains from the conveyor then the adjustment should be slackened off and the chain thoroughly cleaned and protected with lubricant. When the conveyor is re-started the chain should be run for a short time without load, to ensure that the lubricant penetrates to the bearing surfaces.

16.2 Chain adjustment

For optimum performance and correct running all chain systems should be provided with means to compensate for elongation due to wear. As a chain conveyor or elevator is a positive form of drive, no pre-tensioning of the chain is necessary. The only adjustment required is the 'take-up' of the clearances between the pins and bushes in each link; this should be done before the conveyor is run.

The amount of adjustment should allow for the joining up of the chain and elongation due to wear. Wherever possible, the adjusting wheel or track should be set at a convenient position following the drive point. This ensures that the effort required to adjust the chain is minimal. Take-up positions should whenever possible be introduced at positions where the conveyor makes a 180° bend. At these positions, the chain take-up will be equal to twice the adjustment. If a 90° position is unavoidable then track movement, particularly on overhead conveyors, will arise. Where multiple drives are used it is preferable to provide an adjustment for each drive point.

A chain should be maintained in correct adjustment throughout its life. Early adjustment will probably be found necessary due to initial 'bedding in' of the mating components. The amount of adjustment varies according to the length and pitch of the chain and can be estimated as follows:—

$$\text{Adjustment (in)} = \frac{\text{Centre distance (in)} \times \text{factor}}{\text{Chain pitch (in)}}$$

Chain series	*Factor*
3,000 lbf	0·016
6,000/7,500 lbf	0·020
12,000/15,000 lbf	0·030
24,000/30,000 lbf	0·040
36,000/45,000 lbf	0·040
60,000 lbf	0·050
85,000 lbf	0·050

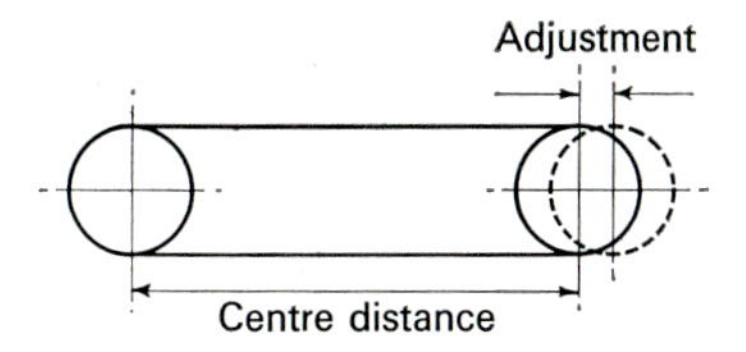

Fig 16.5

EXAMPLE: Conveyor 30 ft centres using 8·0 in pitch. 30,000 lbf chain

$$\text{Adjustment} = \frac{30 \times 12 \times 0{\cdot}040}{8}$$

$$= 1{\cdot}8 \text{ in}$$

When a chain has been adjusted to this extent it will be due for replacement. Where this calculated figure exceeds twice the pitch of the chain, then a minimum adjustment of plus 1·5 pitches, minus 0·5 pitch on the nominal centre distance should be provided. This amount of adjustment will allow the removal of two pitches of chain as wear occurs; the minus adjustment providing sufficient slack for the initial connecting up of the chain.

On dredging feed elevators, where the boot wheel is the adjustable member, provision should be made not only for wheel adjustment but also for the lining of the boot so that the buckets remain at a constant distance from the lining.

The most common type of adjuster in use is the screw take-up type. In this arrangement, shown in Fig 16.6, the tailshaft bearings are mounted in slides and adjusted by adjusting screws. When the desired position is reached, the bearings are locked in position. Automatic adjustment can be achieved by five methods and is generally required on installations where temperature changes are considerable, i.e. drying ovens. The aim of each is to impose a minimum adjusting load or tension in the chain consistent with the take-up of chain slack. Such methods are dead weight, spring take-up, pneumatic or hydraulic counterweight take-up, and chain catenary take-up.

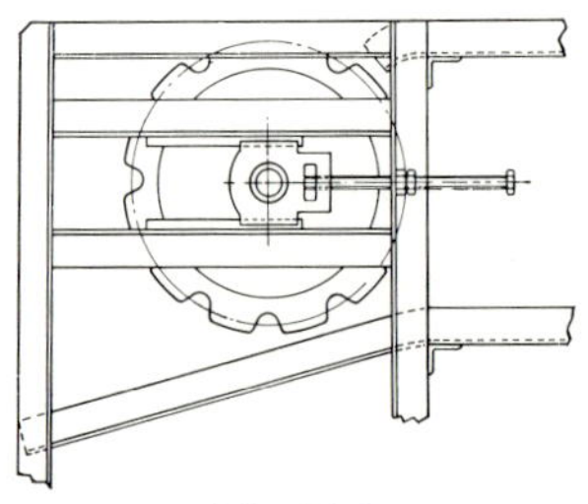

Fig 16.6

16.3 Calculating tension loads

An assessment of the tension load is necessary for automatic take-up systems. In a simple two strand slat conveyor, as represented in Fig 16.7, the tension load T_A applied by the automatic take-up is that required to balance the effects of the unloaded and loaded strands at the take-up position.

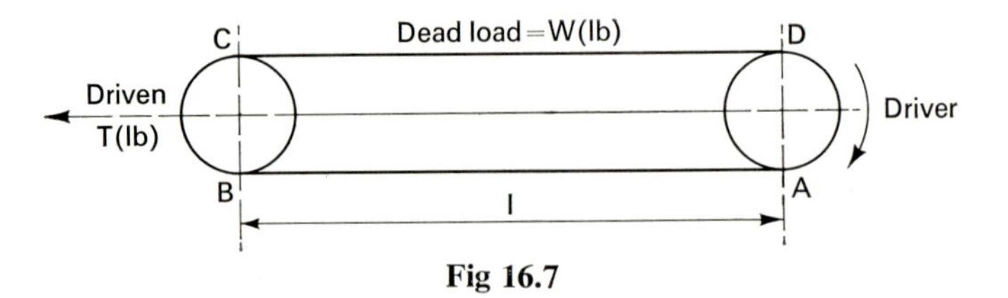

Fig 16.7

The theoretical tension load T_T (lbf) required is given by the following expression:—

$$T_T = \text{Chain pull at B} + \text{chain pull at C}$$

To this should be added the frictional sliding or rolling resistance of the take-up unit. To cater for this it is usual to increase the theoretical tension load by 10%.

Therefore $T_A = 1{\cdot}10$ (chain pull at B + chain pull at C).

Chain lubrication 16.4

Effective lubrication of the chain bearing surfaces is essential to obtain optimum performance in addition to minimising power absorption, rate of wear, liability of corrosion and noise.

The lubricant used for chains must be of a grade capable of reaching the bearing surfaces between the bearing pin and bush, and between the bush and roller, and with adequate body to maintain an oil film over the whole of these surfaces as shown in Fig 16.8. It must also maintain its lubricating properties under operating conditions and be free from corrosive elements.

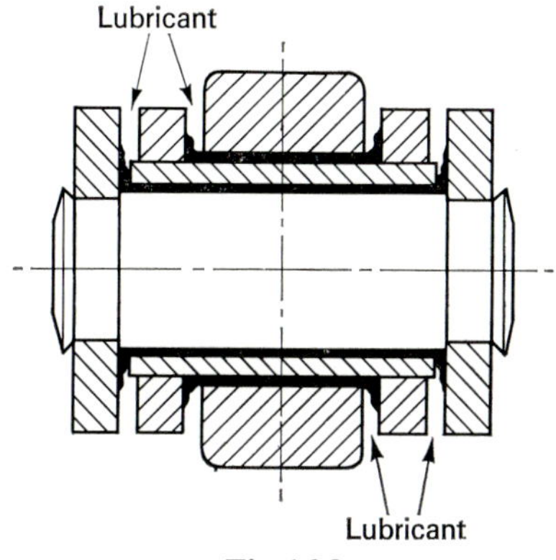

Fig 16.8

In all cases the lubricant should be applied immediately after the chain leaves the driving wheel and with the chain running. This is the point of least tension and the most likely position where the lubricant will reach the rubbing surfaces. Chains can be lubricated automatically with drip feed or oil mist spray lubricators or manually with a brush and lubricant. For normal conditions a good quality mineral base lubricant with a medium viscosity is recommended.

Mineral base lubricants carbonise at about 140°C thus causing a build up of carbon between pin/bush and bush/roller. For temperatures up to 300°C a colloidal graphite lubricant suspended in a volatile carrier should be used. Evaporation of the carrier (usually white spirit) leaves a film of graphite on the bearing surfaces but this will not be retained for a long period and must be re-applied at regular intervals.

Chains operating in abrasive conditions can also be lubricated with a dry lubricant but for extremely abrasive applications grease gun lubricated chain (see Fig 16.9) may be used.

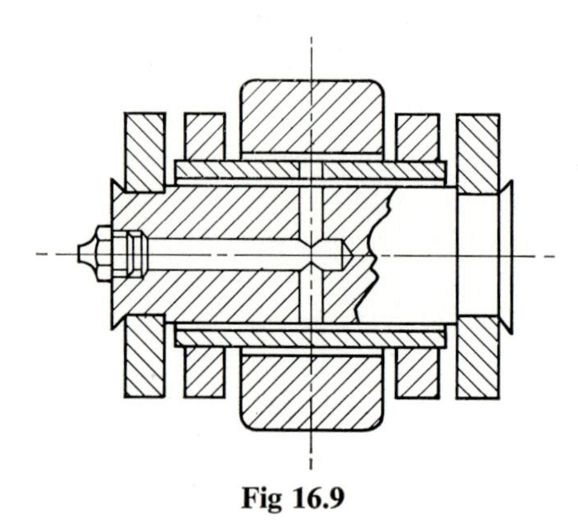

Fig 16.9

For wet conditions, water repellent grease should be used, but this will not penetrate into the chain bearing surfaces and therefore additional application of oil is recommended.

Prevention of contamination of lubricant is usually considered in the context of handling food for human consumption. On such applications, vegetable oils or medicinal paraffin may be acceptable, and will provide satisfactory lubrication when applied by the normal methods. If non lubricated chains are essential, a reduction in chain life must be accepted, and taken into account when selecting the chain. For dairies or similar industries the possibility of bottles being marked by a lubricant is not acceptable, and soluble oil or liquid soap and water are widely used. When designing conveyor systems on which product contamination must occur, every effort should be made to avoid proximity of the chain to the product, as this will obviously assist the provision of satisfactory lubrication without running into contamination difficulties.

More detailed information is available from chain manufacturers.

17

Effects of Adverse Operating Conditions

Introduction 17.1

Generally operating problems are caused by:—

Poor basic design of the conveyor system
Faulty ancillary equipment
Incorrect structural assembly
Lack of correct maintenance, particularly lubrication.

Breakdowns and stoppages caused by any of the foregoing points are often attributed to faulty chains or incorrect chain selection, due to the symptoms first being made evident in poor chain performance or chain failure.

Sequence of fault finding 17.2

Success in fault finding is closely associated with a systematic approach due to the variety of materials handled and associated conveyor designs throughout all the industrial processes. There is no effect without cause; as incidents do not just happen, the cause must be found.

The following sequence is applicable to the majority of cases.

BEFORE SITE INSPECTION

Determine the type of application, material handled, loading details and circuit layout. Obtain full details of the incidents that have arisen in the conveyor equipment and if possible an actual sample of chain and components where applicable.

SITE INSPECTION

Equipment stationary

Examine chain and attachments in detail. Check wheel assemblies and drive arrangements. Examine tracking and structural work for defects. Discuss fault with maintenance fitter or machine operator, as unusual machine performance may assist in indicating the cause.

Equipment running

Observe machine during actual operation, at the same time carrying out further inspection.

The following charts will assist in determining the cause of failure. They are intended as a general guide, but cover a considerable proportion of the situations encountered.

CHART 1
Chain fault finding

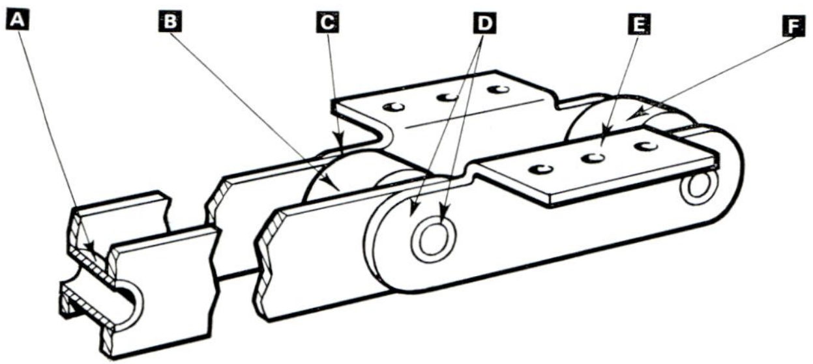

A B C D E F

A

FAULT	CAUSE	REMEDY
Fractured bushes	1. Speed too high for pitch	1. Chain of shorter pitch but equivalent strength
	2. Heavy shock load applied	2. Investigate on-loading in attempt to minimise shock
	3. Corrosion pitting	3. Consider special materials or improve lubrication

B

FAULT	CAUSE	REMEDY
Roller flatting due to skidding	1. Too lightly loaded system	1. Increase load within limits of chain
	2. Heavy load where friction between bush and roller bore overcomes lever effect of friction at roller periphery	2. Increase chain size if no load reduction possible
	3. Excessive lubricant on track	3. Clean and scour track
	4. Canting of chain due to load	4. Strengthen carrying medium

C

FAULT	CAUSE	REMEDY
Tight chain joints	1. Material packed in chain	1. Clean and re-lubricate (1–2)
	2. Material frozen in joints	2. Reduce chain/material contact, run continuously (1–2)
	3. Incorrect lubrication (gummy)	3. Clean and lubricate with correct type of lubricant
	4. Corrosion	4. Investigate cause and consider special materials
	5. Malalignment	5. Check alignment of structure
	6. Plate movement after bush turning in holes	6. Improve pin/bush lubrication

D

FAULT	CAUSE	REMEDY
Fractured plate Fractured bearing pin Elongated holes	Overload above maximum breaking strength	1. Investigate for foreign objects causing jams 2. Protect chain—shear pin device 3. Review loading

E

FAULT	CAUSE	REMEDY
Loose or damaged attachments	1. High unit shock loading	1. Minimise shock by modifying loading sequence
	2. Incorrect slat or carrier assembly	2. Re-align to ensure correct phasing of chains
	3. Twisted chain causing flexure of platform by continual slat or carrier movement	3. Emphasise care at assembly stage in movement of handling lengths

F

FAULT	CAUSE	REMEDY
Excessive roller bore wear	1. High unit load	1. Distribute load—alter pitch
	2. Twisted slats or carriers	2. Rectify and check for flatness
	3. Packing of abrasive particles	3. Minimise chain/material contact—consider chain as pulling medium only
	4. Unsatisfactory roller bore lubrication and corrosion	4. Improve lubrication; change to grease gun design if essential

General fault finding

FAULT	CAUSE	REMEDY
Excessive noise	1. Malalignment of track joints	1. Check alignment of structure
	2. Too little or too much chain slack	2. Adjust correctly
	3. High speed	3. Consider shorter pitch
	4. Chain or wheels worn	4. Replace
	5. Ineffective lubrication	5. Lubricate
	6. Incorrect positioning of guide tracks adjacent to wheels	6. Reposition

FAULT	CAUSE	REMEDY
Uneven running	1. Heavy load and low speed causing rollers to 'stick-slip'	1. See roller flatting Consider additional drive point. Check surge at driver due to inadequate power reserve or shaft/bearing rigidity
	2. Polygonal action of closely spaced wheels in complex circuit	2. Increase wheel centres or reposition wheels
	3. High friction of idler wheels	3. Lubricate correctly or fit low friction bearings
	4. Polygonal action on wheels	4. Introduce wheels with larger number of teeth

FAULT	CAUSE	REMEDY
Chain whip	1. Excessive slack	1. Adjust correctly
	2. Long centres with periodic on-loading of material causing pulsating action	2. Fully guide return strand

CHART 2
Wheel fault finding

FAULT	CAUSE	REMEDY
Chain clings to wheels	1. Incorrect tooth form	1. Replace
	2. Worn tooth form	2. Reverse chain and wheels if possible
	3. Heavy and tacky lubricants	3. Clean and re-lubricate
	4. Stiff chain joints	4. See Chart 1

FAULT	CAUSE	REMEDY
Chain climbs wheels	1. Excessive tooth wear	1. Replace wheels
	2. Build-up of excessive slack	2. Adjust chain correctly
	3. Chain elongation	3. Replace chain
	4. Severe overloads	4. Reduce loading—strengthen carriers—consider special tooth form
	5. Material packing between chain and wheels	5. Relieved teeth
	6. Heavy loads carried under wheel	6. Constrain chain around wheel

CHART 3
Wheel inspection

INSPECTION	POSSIBLE FAULT
Place spirit level on all head and tail shafts and check for accuracy of horizontal setting. Also check visually for squareness of conveyor tracks in plan view.	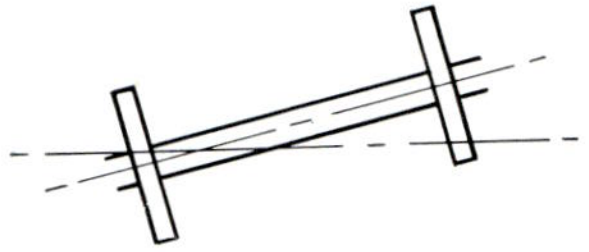
Rotate shaft through 90° and inspect as above.	BENT SHAFT
Release tension weights. Disconnect chain at headwheels, place bar across track to seat in tooth form. Check alignment with spirit level sighting against shaft.	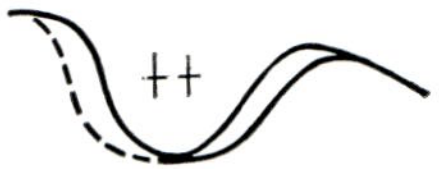 ACCURACY OF LATERAL TOOTH GAP ALIGNMENT
Record rubbing marks made by the chain on each wheel indicating position and severity of wear.	Use 'Engineers Blue' for best results
Ensure one tail wheel of two strand conveyor has free rotation relative to the other.	

CHART 4
Tracking inspection

INSPECTION	POSSIBLE FAULTS
Looking at conveyor from the side, check the flatness of all tracks along conveyor length.	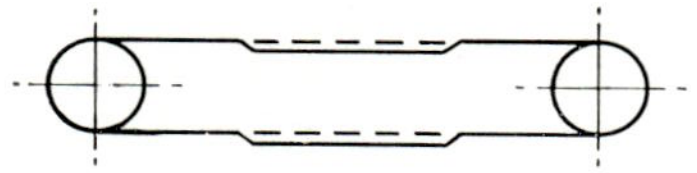 ELEVATION
Check consistency of transverse distance between tracks along length of conveyor.	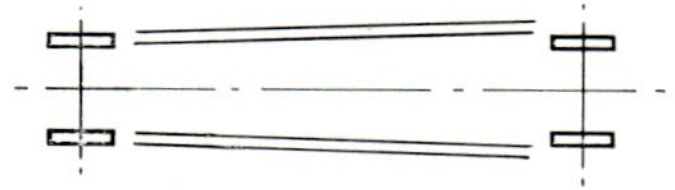 PLAN
Check for side-bow in tracks.	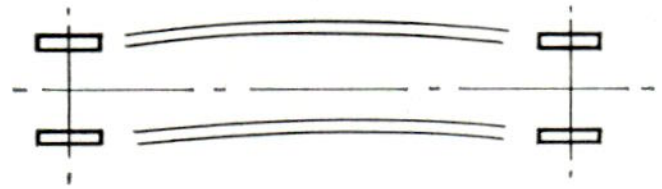 PLAN
Lay straight edge across a pair of tracks and check with spirit level. Repeat at different points along conveyor length.	

INSPECTION	POSSIBLE FAULTS
Lay spirit level across individual tracks and check.	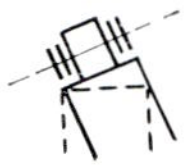
Record transverse centres of wheels and compare with tracking.	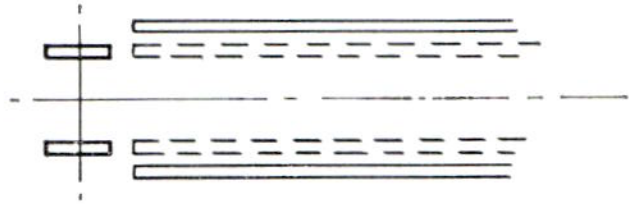 PLA N
Check squareness of a link to carrier or slat.	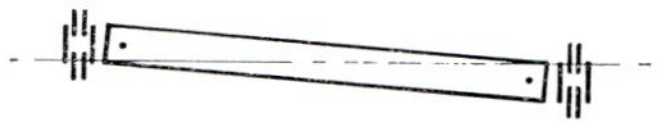
Check height between faces of adjacent slats or carriers on working strands.	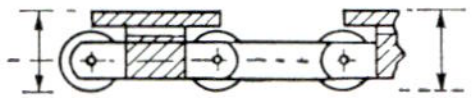
Inspect tracks for any sign of wear or fouling.	

Tables

TABLE 1.1
Precision conveyor chains

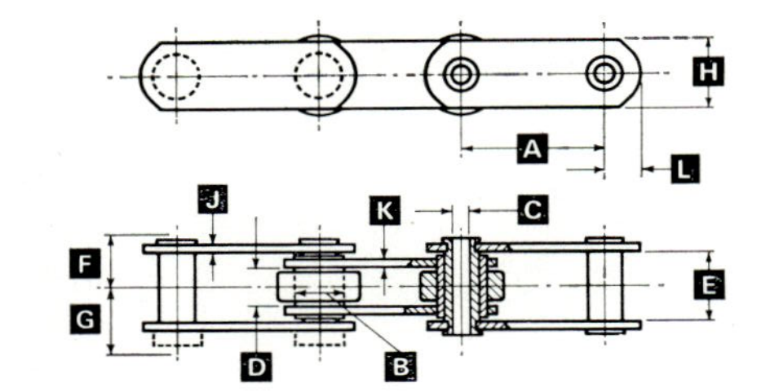

HOLLOW BEARING PIN CHAIN

Chain breaking load	Pitch (min.) A	Pitch (max.) A	Bush diameter B	Hollow pin bore (min.) C	Width between plates Inner D	Width between plates Outer E	Chain track Plain side F	Chain track Fastener side G	Plate Depth H	Plate Thickness Outer J	Plate Thickness Inner K	Plate Head (max.) L
lbf	*in*	*in*	*in*	*in*	*in*	*in*	*in*	*in*	*in*	*in*	*in*	*in*
4,500	1·50	3·00	1·00*	0·26	0·50	0·70	0·64	0·64	0·78	0·07	0·09	0·42
6,000	1·50	6·00	0·71	0·38	0·60	1·00	0·92	0·98	1·00	0·15	0·15	0·58
12,000	2·00	9·00	0·93	0·52	0·75	1·28	1·04	1·13	1·50	0·15	0·20	0·87
24,000	3·50	12·00	1·30	0·77	1·00	1·70	1·35	1·97	2·00	0·20	0·30	1·17
36,000	5·00	18·00	1·50	0·90	1·50	2·33	1·84	2·65	2·40	0·30	0·35	1·37

**Roller diameter—bush chain not available.*

SOLID BEARING PIN CHAIN

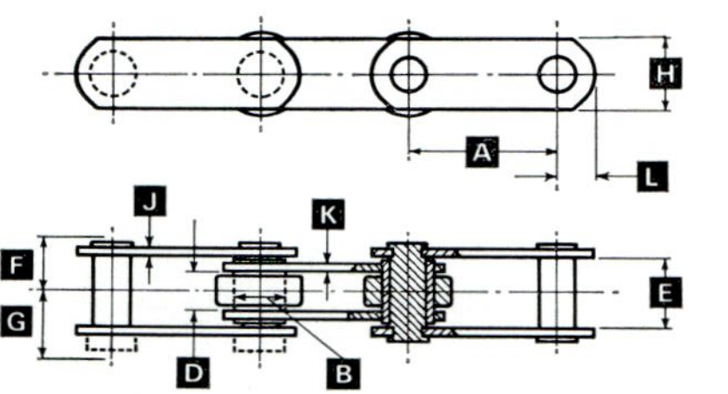

Chain breaking load	Pitch		Bush diameter	Width between plates		Chain track		Plate			
	(min.)	(max.)		Inner	Outer	Plain side	Fastener side	Depth	Thickness		Head (max.)
									Outer	Inner	
	A	A	B	D	E	F	G	H	J	K	L
lbf	*in*	*in*	*in*	*in*	*in*	*in*	*in*	*in*	*in*	*in*	*in*
3,000	1·00	4·50	—	0·46	0·63	0·51	0·56	0·70	0·07	0·07	0·47
7,500	1·50	6·00	0·71	0·60	1·00	0·92	0·98	1·00	0·15	0·15	0·58
15,000	2·00	9·00	0·93	0·75	1·28	1·04	1·13	1·50	0·15	0·20	0·87
30,000	3·50	12·00	1·30	1·00	1·70	1·35	1·97	2·00	0·20	0·30	1·17
45,000	5·00	18·00	1·50	1·50	2·33	1·84	2·65	2·40	0·30	0·35	1·37
60,000	6·00	18·00	1·50	1·50	2·31	1·80	2·10	2.40	0·30	0·35	1·37
85,000	6·00	24·00	1·50	1·50	2·61	2·05	2·45	2·50	0·40	0·50	1·41

TABLE 1.2

Rollers for precision conveyor chains

PLAIN TYPE ROLLERS

Chain breaking load		Minimum chain pitch	Tread diameter	Width	Roller material		
Hollow bearing pin	Solid bearing pin		A	C	Cast iron	Mild steel	Hardened steel
lbf	*lbf*	*in*	*in*	*in*			
Small rollers							
—	3,000	1·00	0·475	0·45			*
6,000	7,500	2·00	1·250	0·55			*
12,000	15,000	3·00	1·875	0·70		*	*
24,000	30,000	4·00	2·625	0·95		*	*
36,000	45,000	5·00	3·500	1·45	*		*
—	60,000	6·00	3·500	1·45	*		*
—	85,000	6·00	3·500	1·45	*		*
Large rollers							
—	3,000	1·50	1·00	0·45			*
36,000	45,000	8·00	5·00	1·45	*		*
—	60,000	8·00	5·00	1·45	*		*
—	85,000	8·00	5·00	1·45	*		*

C

A

FLANGED TYPE ROLLERS

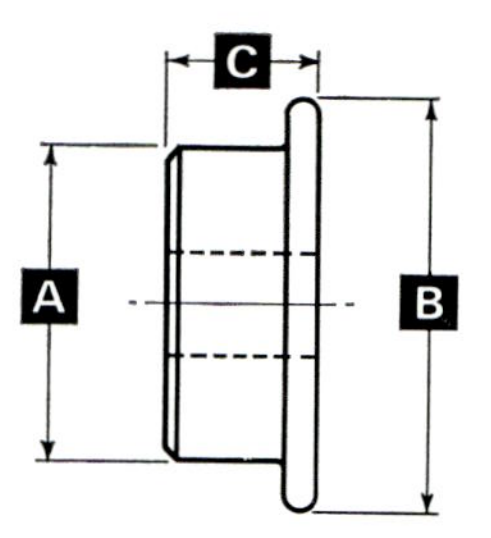

Chain breaking load		Minimum chain pitch	Tread diameter	Flange diameter	Width	Roller material		
Hollow bearing pin	Solid bearing pin		A	B	C	Cast iron	Mild steel	Hardened steel
lbf	*lbf*	*in*	*in*	*in*	*in*			
		Small rollers						
6,000	7,500	2·50	1·250	1·625	0·55		*	*
12,000	15,000	3·50	1·875	2·375	0·70		*	*
24,000	30,000	4·50	2·625	3·375	0·95		*	*
36,000	45,000	6·00	3·500	4·500	1·45	*		*
—	60,000	6·00	3·500	4·500	1·45	*		*
—	85,000	7·00	3·500	4·500	1·45	*		*
		Large rollers						
36,000	45,000	9·00	5·00	6·00	1·45	*		*
—	60,000	9·00	5·00	6·00	1·45	*		*
—	85,000	9·00	5·00	6·00	1·45	*		*

TABLE 2.1 Factor f_1

Conveyor inclination	Factor	Conveyor inclination	Factor	Conveyor inclination	Factor
Horizontal to 5°	0.22	Over 10° to 20°	0·42	Over 40° to 60°	0·86
Over 5° to 10°	0·30	Over 20° to 40°	0·64	Over 60° to vertical	1·0

TABLE 2.2 Factor f_4

Resistance factor for various material sliding on steel

Material	Conveyor inclination									
	0°	5°	10°	15°	20°	30°	40°	50°	60°	70°
Brass (unlubricated)	0·20	0·29	0·37	0·45	0·53	0·67	0·80	0·89	0·97	1·01
Wood Cases	0·30	0·39	0·47	0·55	0·62	0·76	0·87	0·96	1·02	1·05
Grain, Anthracite	0·40	0·49	0·57	0·65	0·72	0·85	0·95	1·02	1·06	1·08
Bituminous Coal, Dry Ashes	0·50	0·59	0·67	0·74	0·81	0·93	1·02	1·09	1·12	
Stone	0·60	0·68	0·76	0·84	0·91	1·02	1·10	1·16	1·17	
Gravel, Dry Sand, Moist Ashes	0·70	0·78	0·86	0·94	0·99	1·10	1·18	1·21	1·22	
Wet Sand	0·80	0·88	0·96	1·03	1·09	1·19	1·26	1·28		

TABLE 2.3 Safety factors

Condition	Factor
Clean, regular lubrication	8
Moderately clean, occasional lubrication	10
Dirty, no lubrication	12
Abrasive	14

TABLE 2.4 Roller loading

Roller material	Rubbing speed V (ft/min)		Maximum pressure P_R(lbf/in^2)	
	Good conditions	Average conditions	Good conditions	Average conditions
Steel	5–30 Over 30	5–50 Over 50	300 Use $P_RV=9{,}000$	60 Use $P_RV=3{,}000$
Cast iron	5–30 Over 30	5–50 Over 50	500 Use $P_RV=15{,}000$	90 Use $P_RV=4{,}500$
Hardened steel	5–30 Over 30	5–50 Over 50	1500 Use $P_RV=45{,}000$	250 Use $P_RV=12{,}500$

TABLE 2.5 Roller loads

Chain breaking load	Roller diameter	Maximum roller load			Maximum chain speed
		Steel roller	Cast iron roller	Hardened steel roller	
lbf	*in*	*lbf*	*lbf*	*lbf*	*ft/min*
3,000	0·475	—	—	40	70
	1·0	—	—	40	100
4,500	1·0	—	—	55	100
6,000/7,500	1·25	25	—	90	100
12,000/15,000	1·875	40	—	160	100
24,000/30,000	2·625	70	—	300	100
36,000/45,000 } 60,000/85,000 }	3·5	—	200	540	100
	5·0	—	200	540	150

The above roller loadings are based on the maximum chain speeds given and on the chain being regularly lubricated, running in clean conditions and absence of abrasion.

TABLE 2.6 Factor f_2

Chain breaking load	Lubrication	Conveyor inclination											
		0°	5°	10°	15°	20°	30°	40°	50°	60°	70°	80°	90°
lbf													
3,000*	Regular	0·17	0·26	0·34	0·42	0·50	0·65	0·77	0·87	0·95	1·0	1·0	1·0
	Occasional	0·19	0·28	0·36	0·44	0·52	0·66	0·79	0·89	0·96	1·0	1·0	1·0
	None	0·21	0·30	0·38	0·46	0·54	0·68	0·80	0·90	0·97	1·0	1·0	1·0
6,000/7,500	Regular	0·18	0·27	0·35	0·43	0·51	0·66	0·78	0·88	0·96	1·0	1·0	1·0
	Occasional	0·21	0·30	0·38	0·46	0·54	0·68	0·80	0·90	0·97	1·0	1·0	1·0
	None	0·24	0·33	0·41	0·49	0·57	0·71	0·81	0·92	0·99	1·0	1·0	1·0
12,000/15,000	Regular	0·14	0·23	0·31	0·39	0·47	0·62	0·75	0·85	0·93	0·99	1·0	1·0
	Occasional	0·17	0·26	0·34	0·42	0·50	0·65	0·77	0·87	0·95	1·0	1·0	1·0
	None	0·19	0·28	0·36	0·44	0·52	0·66	0·79	0·89	0·96	1·0	1·0	1·0
24,000/30,000	Regular	0·12	0·21	0·30	0·37	0·45	0·60	0·73	0·84	0·93	0·98	1·0	1·0
	Occasional	0·15	0·24	0·32	0·40	0·48	0·63	0·76	0·86	0·94	0·99	1·0	1·0
	None	0·17	0·26	0·34	0·42	0·50	0·65	0·77	0·87	0·95	1·0	1·0	1·0
36,000/45,000†	Regular	0·10	0·19	0·27	0·35	0·43	0·59	0·72	0·83	0·92	0·97	1·0	1·0
60,000 †	Occasional	0·12	0·21	0·30	0·37	0·45	0·60	0·73	0·84	0·93	0·98	1·0	1·0
85,000 †	None	0·14	0·23	0·31	0·39	0·47	0·62	0·75	0·85	0·94	0·99	1·0	1·0

* *Based on 1·0 in diameter rollers*

† *Based on 3·5 in diameter rollers*

TABLE 2.7 Factor f_3

Chain breaking load	Lubrication	Conveyor inclination															
		0°	5°	6°	7°	8°	9°	10°	15°	20°	30°	40°	50°	60°	70°	80°	90°
lbf	Regular	0·17	0·08	0·06	0·04	0·03	0·01	0	0	0	0	0	0	0	0	0	0
3,000*	Occasional	0·19	0·10	0·08	0·07	0·05	0·03	0·01	0	0	0	0	0	0	0	0	0
	None	0·21	0·12	0·10	0·09	0·07	0·05	0·03	0	0	0	0	0	0	0	0	0
	Regular	0·18	0·09	0·07	0·06	0·04	0·02	0	0	0	0	0	0	0	0	0	0
6,000/7,500	Occasional	0·21	0·12	0·10	0·09	0·07	0·05	0·03	0	0	0	0	0	0	0	0	0
	None	0·24	0·15	0·13	0·12	0·10	0·08	0·06	0·03	0	0	0	0	0	0	0	0
	Regular	0·14	0·05	0·03	0·02	0	0	0	0	0	0	0	0	0	0	0	0
12,000/15,000	Occasional	0·17	0·08	0·06	0·04	0·03	0·01	0	0	0	0	0	0	0	0	0	0
	None	0·19	0·10	0·08	0·07	0·05	0·03	0·01	0	0	0	0	0	0	0	0	0
	Regular	0·12	0·03	0·01	0	0	0	0	0	0	0	0	0	0	0	0	0
24,000/30,000	Occasional	0·15	0·06	0·04	0·03	0·01	0	0	0	0	0	0	0	0	0	0	0
	None	0·17	0·08	0·06	0·04	0·03	0·01	0	0	0	0	0	0	0	0	0	0
36,000/45,000†	Regular	0·10	0·01	0	0	0	0	0	0	0	0	0	0	0	0	0	0
60,000 †	Occasional	0·12	0·03	0·01	0	0	0	0	0	0	0	0	0	0	0	0	0
85,000 †	None	0·14	0·05	0·03	0·02	0	0	0	0	0	0	0	0	0	0	0	0

**Based on 1·0 in diameter rollers*

†Based on 3·5 in diameter rollers

TABLE 2.8

x	e^x	x	e^x	x	e^x
0·02	1·0202	0·25	1·2840	0·45	1·5683
0·04	1·0408	0·26	1·2969	0·46	1·5841
0·06	1·0618	0·27	1·3100	0·47	1·6000
0·08	1·0833	0·28	1·3231	0·48	1·6161
		0·29	1·3364	0·49	1·6323
0·10	1·1052	0·30	1·3499	0·50	1·6487
0·11	1·1163	0·31	1·3634	0·56	1·8221
0·12	1·1275	0·32	1·3771	0·57	2·0138
0·13	1·1388	0·33	1·3910	0·58	2·2255
0·14	1·1505	0·34	1·4050	0·59	2·4596
0·15	1·1618	0·35	1·4191	1·0	2·7183
0·16	1·1735	0·36	1·4333	1·1	3·0042
0·17	1·1835	0·37	1·4477	1·2	3·3201
0·18	1·1972	0·38	1·4623	1·3	3·6693
0·19	1·2092	0·39	1·4770	1·4	4·0552
0·20	1·2214	0·40	1·4918	1·5	4·4817
0·21	1·2337	0·41	1·5068	1·6	4·9530
0·22	1·2461	0·42	1·5220	1·7	5·4739
0·23	1·2586	0·43	1·5373	1·8	6·0497
0·24	1·2712	0·44	1·5527	1·9	6·6859
				2·0	7·3891

TABLE 2.9 Factor f_5

Chain breaking load	Lubrication	Conveyor inclination											
		0°	5°	10°	15°	20°	30°	40°	50°	60°	70°	80°	90°
lbf													
3,000*	Regular	0·17	0·17	0·17	0·16	0·16	0·15	0·13	0·10	0·08	0·06	0·02	0
	Occasional	0·19	0·19	0·19	0·18	0·18	0·16	0·15	0·13	0·10	0·07	0·03	0
	None	0·21	0·21	0·21	0·20	0·20	0·18	0·16	0·14	0·11	0·07	0·03	0
6,000/7,500	Regular	0·18	0·18	0·17	0·17	0·17	0·16	0·14	0·11	0·09	0·06	0·02	0
	Occasional	0·21	0·21	0·21	0·20	0·20	0·18	0·16	0·14	0·11	0·07	0·03	0
	None	0·24	0·25	0·24	0·23	0·22	0·21	0·17	0·16	0·12	0·07	0·03	0
12,000/15,000	Regular	0·14	0·14	0·14	0·14	0·13	0·12	0·11	0·08	0·06	0·05	0·02	0
	Occasional	0·17	0·17	0·17	0·16	0·16	0·15	0·13	0·10	0·08	0·06	0·02	0
	None	0·19	0·19	0·19	0·18	0·18	0·16	0·15	0·13	0·10	0·07	0·03	0
24,000/30,000	Regular	0·12	0·12	0·12	0·11	0·11	0·10	0·09	0·07	0·06	0·04	0·02	0
	Occasional	0·15	0·15	0·15	0·15	0·14	0·13	0·12	0·10	0·08	0·05	0·02	0
	None	0·17	0·17	0·17	0·16	0·16	0·15	0·13	0·10	0·08	0·06	0·02	0
36,000/45,000†	Regular	0·10	0·10	0·10	0·10	0·09	0·09	0·07	0·06	0·05	0·03	0·01	0
60,000 †	Occasional	0·12	0·12	0·12	0·11	0·11	0·10	0·09	0·07	0·06	0·04	0·02	0
85,000 †	None	0·14	0·14	0·14	0·14	0·13	0·12	0·11	0·08	0·06	0·05	0·02	0

* *Based on 1·0 in diameter rollers*

† *Based on 3·5 in diameter rollers*

TABLE 10.1

Horizontal bend factors for Vertichain										
Bend lap angle (Degrees)	10	20	30	40	45	50	60	70	80	90
Horizontal plane roller	1·014	1·028	1·043	1·057	1·065	1·072	1·087	1·103	1·118	1·134
Bend lap angle	100	110	120	130	140	150	160	170	180	
Horizontal plane roller	1·150	1·166	1·182	1·199	1·216	1·233	1·250	1·268	1·285	

For any other angle the drag factor is given by the following expression:—

$e^{\mu\theta}$ where $e=2{\cdot}718$
$\theta=$ Bend angle (Radians)
$\mu=0{\cdot}08$

TABLE 10.2

Incline (Degrees)	10	15	20	22·5	25	30	35	40	45
μ Cos θ + Sin θ	0·208	0·293	0·375	0·42	0·454	0·53	0·602	0·67	0·732
μ Cos θ − Sin θ	−0·14	−0·224	−0·31	−0·35	−0·391	−0·47	−0·545	−0·62	−0·682

where $\mu=0{\cdot}035$

TABLE 10.3

Vertical bend drag factors										
Bend lap angle (Degrees)	10	20	30	40	45	50	60	70	80	90
Vertical plane roller	1·006	1·012	1·019	1·025	1·028	1·031	1·037	1·044	1·050	1·057
Bend lap angle	100	110	120	130	140	150	160	170	180	
Vertical plane roller	1·063	1·069	1·076	1·082	1·089	1·095	1·103	1·110	1·116	

For any other angle the drag factor is given by

$e^{\mu\theta}$ where $e=2{\cdot}718$
and $\theta=$ Bend angle (Radians)
$\mu=0{\cdot}035$

TABLE 12.1 Agricultural chain

BS chain number	Pitch	Roller diameter	Width between inner plates	Breaking load
	in	*in*	*in*	*lbf*
S.32	1·15	0·45	0·625	1,800
S.42	1·375	0·562	0·750	6,000
S.52	1·50	0·60	0·875	4,000
S.55	1·63	0·70	0·875	4,000
S.45	1·63	0·60	0·875	4,000
S.62	1·65	0·75	1·00	6,000
S.77	2·297	0·719	0·875	10,000
S.88	2·609	0·90	1·125	10,000

TABLE 13.1

No. of teeth	Unit pitch circle diameter	No. of teeth	Unit pitch circle diameter	No. of teeth	Unit pitch circle diameter
6	2·000	21	6·709	36	11·474
7	2·305	22	7·027	37	11·792
8	2·613	23	7·344	38	12·110
9	2·924	24	7·661	39	12·428
10	3·236	25	7·979	40	12·746
11	3·549	26	8·296	41	13·063
12	3·864	27	8·614	42	13·382
13	4·179	28	8·931	43	13·700
14	4·494	29	9·249	44	14·018
15	4·810	30	9·567	45	14·336
16	5·126	31	9·885	46	14·654
17	5·442	32	10·202	47	14·972
18	5·759	33	10·520	48	15·290
19	6·076	34	10·838	49	15·608
20	6·392	35	11·156	50	15·926

TABLE 13.2

Application	Normal range of wheels		
	No. of teeth		
Slat, Bar, Steel Apron, Wire Mesh or similar Conveyors	8–12		
Finger Tray, Soft Fruit and similar Elevators	8–12		
Cask, Package and similar Elevators	8–12		
Swing Tray Elevators	16–24		
Ore Feeder Conveyors	6–8		
Scraper Conveyors	8–10		
Box Scraper Conveyors	8–10		
	Normal minimum No. of teeth in wheel		
Bucket Elevators	Head	Boot	Deflector
Spaced Bucket			
High speed; vertical (one or two chains)	14	11	—
Medium speed; inclined (one or two chains)	14	11	—
Slow speed; vertical (two chains)	12	11	9
Continuous Bucket			
Medium speed; vertical or inclined (one or two chains)	8	8	—
Slow speed; vertical or inclined (two chains)	6	6	—
	Driver	Top corner	Follower
Gravity Bucket Conveyor/Elevators	12	12	8

TABLE 13.3

Number of teeth	8	10	12	14	16	18	20	22	24
Dimension (x)	0·099	0·079	0·066	0·057	0·049	0·044	0·039	0·036	0·033

TABLE 13.4

Number of teeth	8	10	12	14	16	20
% Speed variation	7·6	4·9	3·4	2·5	1·9	1·2

Appendices

Appendix 1

MATERIALS COMMONLY HANDLED
(Approximate Values)

Material	Average weight (lb/cu ft)	Angle of repose (Degrees)	Maximum conveyor inclination (Degrees)‡
Alum (Lumpy)	55		
Anthracite(Broken)*	55	38/45	
Ashes (Dry)*	40	40	22/25
Ashes (Wet)*	50		
Bagasse	7		
Barley	40		
Beans	52		8
Bones (Crushed)	40		
Books	40		
Borax (Powdered)	53		
Bran	21		
Carbon Black (Pellets)	25	50	
Cement Clinker*	80	30/40	18/20
Cement*	70	50	20/23
Chalk (Crushed)	90		
Chalk (Powdered)	30		
Charcoal*	22		
Clay (Dry fines)*	95		
Clay (Damp)	120		
Coal (Pulverised)	32		
Coal (Run of mine)	50	38	18
Coffee Beans	28		
Coffee (Ground)	28		
Coke (Broken)*	35	30/40	20
Cork	14		
Corn (Grits)	45	32/45	
Corn Meal	40		
Cotton Seed (Dry)	25		
Dolomite(Crushed)*	95		
Earth (Loose)*	80		
Feldspar (Ground)*	75		

‡This applies to conveyors without pusher flight attachments
*Abrasive material

Material	Average weight (lb/cu ft)	Angle of repose (Degrees)	Maximum conveyor inclination (Degrees)‡
Flour	35		
Flint*	162		
Fuller's Earth (Raw)	40		
Glass Gullet*	90		20
Glass Plate*	172		
Grain	35		15/18
Granite*	167		
Gravel*	110	30	12/15
Gypsum (½″ screening)*	82	40	21
Gypsum (1½″–3″ lumps)	70/80	30	15
Hay (Bales)	23		
Hops, Spent (Dry)	35		
Hops (Wet)	50		
Ice	59		
Ice (Crushed)	40		
Iron (Cast, Pig)*	450		
Iron Ore (Crushed)*	140		20
Lime (Ground)	64		
Lime (Dehydrated)	36	38	
Limestone (Crushed)*	90	38	18
Limestone (Pulverised)*	70		
Malt, Whole (Dry)	25		
Malt (Wet)	63		
Masonry (Rubble)*	140		
Oats	29		
Phosphate Rock (Broken Dry)	80	40	12/15
Quartz (Crushed)*	100		
Rubber (Ground)	23		
Rye	45		
Salt (Coarse)*	47		
Salt (Cake, Dry, Coarse)*	85	36	21
Salt (Dry, fines)*	75	25	11
Sand (Dry)*	100		15
Sand (Damp)*	120		20
Sand (Foundry)*	100		24
Sand (Silica)*	100	42/44	
Sawdust	15		
Shale (Crushed)*	90	39	22

‡This applies to conveyors without pusher flight attachments
*Abrasive material

Material	Average weight (lb/cu ft)	Angle of repose (Degrees)	Maximum conveyor inclination (Degrees)‡
Slag (Furnace, crushed)*	80	25	10
Soda Ash (Heavy)*	60		
Soda Ash (Fluffed)*	27		
Soya beans*	48		
Starch	45		
Stone (Crushed)*	100		18
Steel Chips & Turnings (Crushed)*	80/150		
Steel Chips & Turnings (Uncrushed)*	25/85		
Sugar (Raw)	60		
Sugar (Refined)	55	38/45	
Sulphur	90/125		
Tanbark (Ground)	55		
Water	62		
Wheat	47		
Woodchips	25		27
Zinc Ore (Crushed)	160	38	22

‡This applies to conveyors without pusher flight attachments
*Abrasive material

Appendix 2

SHAFT DIAMETERS

The table opposite gives a method for determining shaft diameters based on the use of mild steel bar (28/32 tons tensile).

Shafts subjected to twisting moments only can be determined from columns 1 and 3; shafts subjected to both bending and twisting moments must be selected from column 2 and appropriate column 'K'.

When selecting shafts subject to twisting moments only: first determine the ratio hp/rpm, and then select shaft diameter from column 3, making any interpolations which may prove necessary.

When selecting shafts subject to both bending and twisting: first determine the twisting and bending moments to which the shaft is subjected. Then determine

$$\text{ratio 'K'} = \frac{\text{Bending moment}}{\text{Twisting moment}}$$

Having determined ratio 'K', select shaft diameter by reference to column 2 (twisting moment) and appropriate column 'K'.

For two strand chain systems with wheels mounted close to the bearings, it can generally be assumed that ratio 'K' will not exceed 1·0.

If the slat conveyor shown on page 30 has a headshaft speed of 7·5 rpm then hp/rpm=0·19. If all wheels are placed close to the bearings ratio 'K' can be taken as 1·0, and a standard shaft diameter of 2·625 in should be used.

hp/ rpm	Twisting moment T	Shaft diameters				
		For twisting only	For bending and twisting			
			K=0·5	K=0·75	K=1·0	K=1·5
	lbf.in	in	in	in	in	in
0·012	756	0·753	0·881	0·948	1·009	1·121
0·025	1,575	0·963	1·126	1·213	1·290	1·434
0·050	3,151	1·213	1·428	1·528	1·625	1·806
0·075	4,726	1·389	1·624	1·750	1·862	2·070
0·100	6,302	1·529	1·788	1·926	2·050	2·277
0·150	9,454	1·750	2·047	2·202	2·345	2·607
0·200	12,605	1·926	2·255	2·425	2·580	2·870
0·250	15,756	2·075	2·427	2·615	2·780	3·090
0·300	18,908	2·205	2·580	2·780	2·950	3·280
0·350	22,059	2·321	2·710	2·920	3·110	3·460
0·400	25,210	2·428	2·840	3·060	3·250	3·620
0·450	28,361	2·524	2·950	3·180	3·380	3·760
0·500	31,513	2·614	3·060	3·290	3·500	3·900
0·600	37,815	2·777	3·250	3·500	3·730	4·140
0·700	44,118	2·925	3·420	3·680	3·920	4·360
0·800	50,420	3·060	3·580	3·850	4·100	4·560
0·900	56,723	3·180	3·720	4·010	4·260	4·740
1·000	63,025	3·290	3·850	4·140	4·410	4·900
1·250	78,782	3·550	4·150	4·470	4·760	5·290
1·500	94,538	3·770	4·410	4·750	5·050	5·620
1·750	110,294	3·970	4·640	5·000	5·320	5·920
2·000	126,051	4·150	4·860	5·230	5·560	6·180
2·250	141,807	4·320	5·050	5·440	5·790	6·440
2·500	157,563	4·470	5·230	5·630	6·000	6·660
2·750	173,320	4·610	5·390	5·810	6·180	6·870
3·000	189,076	4·750	5·560	5·980	6·370	7·080
3·250	204,833	4·880	5·710	6·150	6·540	7·270
3·500	220,589	5·000	5·850	6·300	6·700	7·450
3·750	236,345	5·120	5·990	6·450	6·860	7·630
4·000	252,102	5·230	6·120	6·590	7·010	7·800
4·250	267,858	5·340	6·250	6·730	7·160	7·960
4·500	283,614	5·440	6·360	6·850	7·300	8·110
4·750	299,371	5·540	6·480	6·980	7·430	8·260
5·000	315,127	5·630	6·590	7·100	7·550	8·390
5·500	346,640	5·820	6·810	7·330	7·800	8·670
6·000	378,152	5·990	7·010	7·550	8·030	8·930
6·500	409,665	6·150	7·200	7·750	8·250	9·170
7·000	441,178	6·300	7·370	7·940	8·450	9·390

Appendix 3

METRIC CONVERSION FACTORS

	To convert	Into	Multiply by
Length	inches (in)	millimetres (mm)	25·4
	feet (ft)	metres (m)	0·3048
Area	square inches (in^2)	square millimetres (mm^2)	645
	square feet (ft^2)	square metres (m^2)	0·0929
Volume	cubic inches (in^3)	cubic millimetres (mm^3)	16,400
	cubic feet (ft^3)	cubic metres (m^3)	0·0283
	pints	litres	0·568
	gallons	litres	4·55
Mass	pounds (lb)	kilogrammes (kg)	0·454
	tons (ton)	kilogrammes (kg)	1,020
Density	pounds per cubic inch (lb/in^3)	megagrammes per cubic metre (Mg/m^3)	27·7
	pounds per cubic foot (lb/ft^3)	kilogrammes per cubic metre (kg/m^3)	16·0
Mass flow rate	pounds per hour (lb/h)	kilogrammes per second (kg/s)	0·000 126
Volume flow rate	cubic feet per second (ft^3/s)	cubic metres per second (m^3/s)	0·0283
	gallons per minute (gal/min)	cubic metres per second (m^3/s)	0·000 075 8
Power	horsepower (hp)	watts (W)	746

Energy	foot pound force (ft lbf)	joules (J)	1·36
	kilowatt hour (kWh)	megajoules (MJ)	3·6
Force (weight)	pounds force (lbf)	newtons (N)	4·45
	kilogrammes force (kgf)	newtons (N)	9·81
	tons force (tonf)	kilonewtons (kN)	9·96
Torque or moment of force	pounds force feet (lbf ft)	newton metres (Nm)	1·36
	pounds force inches (lbf in)	millinewton metres (mNm)	113
	tons force feet (tonf ft)	kilonewton metres (kNm)	3·04
Pressure and stress	pounds force per square inch (lbf/in^2)	newtons per square metre (N/m^2)	6,890
	pounds force per square foot (lbf/ft^2)	newtons per square metre (N/m^2)	47·9
	tons force per square inch (tonf/in^2)	meganewtons per square metre (MN/m^2)	15·4
	tons force per square foot (tonf/ft^2)	kilonewtons per square metre (kN/m^2)	107
Velocity	feet per second (ft/s)	metres per second (m/s)	0·3048
	feet per minute (ft/min)	millimetres per second (mm/s)	5·08
	miles per hour (mile/h)	metres per second (m/s)	0·447
Acceleration	feet per second^2 (ft/s^2)	metres per second^2 (m/s^2)	0·3048
Momentum	pounds feet per second (lb ft/s)	kilogrammes metres per second (kg m/s)	0·138

Appendix 4

Miscellaneous technical data

CONVERSION TABLE

Inches to Millimetres

in	0·000	0·001	0·002	0·003	0·004	0·005	0·006	0·007	0·008	0·009
	mm	mm	mm	mm	mm	mm	mm	mm	mm	mm
0·00	0·0000	0·0254	0·0508	0·0762	0·1016	0·1270	0·1524	0·1778	0·2032	0·2286
0·01	0·2540	0·2794	0·3048	0·3302	0·3556	0·3810	0·4064	0·4318	0·4572	0·4826
0·02	0·5080	0·5334	0·5588	0·5842	0·6096	0·6350	0·6604	0·6858	0·7112	0·7366
0·03	0·7620	0·7874	0·8128	0·8382	0·8636	0·8890	0·9144	0·9398	0·9652	0·9906
0·04	1·0160	1·0414	1·0668	1·0922	1·1176	1·1430	1·1684	1·1938	1·2192	1·2446
0·05	1·2700	1·2954	1·3208	1·3462	1·3716	1·3970	1·4224	1·4478	1·4732	1·4986
0·06	1·5240	1·5494	1·5748	1·6002	1·6256	1·6510	1·6764	1·7018	1·7272	1·7526
0·07	1·7780	1·8034	1·8288	1·8542	1·8796	1·9050	1·9304	1·9558	1·9812	2·0066
0·08	2·0320	2·0574	2·0828	2·1082	2·1336	2·1590	2·1844	2·2098	2·2352	2·2606
0·09	2·2860	2·3114	2·3368	2·3622	2·3876	2·4130	2·4384	2·4638	2·4892	2·5146
0·10	2·5400	2·5654	2·5908	2·6162	2·6416	2·6670	2·6924	2·7178	2·7432	2·7686
0·11	2·7940	2·8194	2·8448	2·8702	2·8956	2·9210	2·9464	2·9718	2·9972	3·0226
0·12	3·0480	3·0734	3·0988	3·1242	3·1496	3·1750	3·2004	3·2258	3·2512	3·2766
0·13	3·3020	3·3274	3·3528	3·3782	3·4036	3·4290	3·4544	3·4798	3·5052	3·5306
0·14	3·5560	3·5814	3·6068	3·6322	3·6576	3·6830	3·7084	3·7338	3·7592	3·7846
0·15	3·8100	3·8354	3·8608	3·8862	3·9116	3·9370	3·9624	3·9878	4·0132	4·0386

0·16	4·0640	4·0894	4·1148	4·1402	4·1656	4·1910	4·2164	4·2418	4·2672	4·2926
0·17	4·3180	4·3434	4·3688	4·3942	4·4196	4·4450	4·4704	4·4958	4·5212	4·5466
0·18	4·5720	4·5974	4·6228	4·6482	4·6736	4·6990	4·7244	4·7498	4·7752	4·8006
0·19	4·8260	4·8514	4·8768	4·9022	4·9276	4·9530	4·9784	5·0038	5·0292	5·0546
0·20	5·0800	5·1054	5·1308	5·1562	5·1816	5·2070	5·2324	5·2578	5·2832	5·3086
0·21	5·3340	5·3594	5·3848	5·4102	5·4356	5·4610	5·4864	5·5118	5·5372	5·5626
0·22	5·5880	5·6134	5·6388	5·6642	5·6896	5·7150	5·7404	5·7658	5·7912	5·8166
0·23	5·8420	5·8674	5·8928	5·9182	5·9436	5·9690	5·9944	6·0198	6·0452	6·0706
0·24	6·0960	6·1214	6·1468	6·1722	6·1976	6·2230	6·2484	6·2738	6·2992	6·3246
0·25	6·3500	6·3754	6·4008	6·4262	6·4516	6·4770	6·5024	6·5278	6·5532	6·5786
0·26	6·6040	6·6294	6·6548	6·6802	6·7056	6·7310	6·7564	6·7818	6·8072	6·8326
0·27	6·8580	6·8834	6·9088	6·9342	6·9596	6·9850	7·0104	7·0358	7·0612	7·0866
0·28	7·1120	7·1374	7·1628	7·1882	7·2136	7·2390	7·2644	7·2898	7·3152	7·3406
0·29	7·3660	7·3914	7·4168	7·4422	7·4676	7·4930	7·5184	7·5438	7·5692	7·5946
0·30	7·6200	7·6454	7·6708	7·6962	7·7216	7·7470	7·7724	7·7978	7·8232	7·8486
0·31	7·8740	7·8994	7·9248	7·9502	7·9756	8·0010	8·0264	8·0518	8·0772	8·1026
0·32	8·1280	8·1534	8·1788	8·2042	8·2296	8·2550	8·2804	8·3058	8·3312	8·3566
0·33	8·3820	8·4074	8·4328	8·4582	8·4836	8·5090	8·5344	8·5598	8·5852	8·6106
0·34	8·6360	8·6614	8·6868	8·7122	8·7376	8·7630	8·7884	8·8138	8·8392	8·8646
0·35	8·8900	8·9154	8·9408	8·9662	8·9916	9·0170	9·0424	9·0678	9·0932	9·1186
0·36	9·1440	9·1694	9·1948	9·2202	9·2456	9·2710	9·2964	9·3218	9·3472	9·3726
0·37	9·3980	9·4234	9·4488	9·4742	9·4996	9·5250	9·5504	9·5758	9·6012	9·6266
0·38	9·6520	9·6774	9·7028	9·7282	9·7536	9·7790	9·8044	9·8298	9·8552	9·8806
0·39	9·9060	9·9314	9·9568	9·9822	10·0076	10·0330	10·0584	10·0838	10·1092	10·1346
0·40	10·1600	10·1854	10·2108	10·2362	10·2616	10·2870	10·3124	10·3378	10·3632	10·3886

in	0·000	0·001	0·002	0·003	0·004	0·005	0·006	0·007	0·008	0·009
	mm	mm	mm	mm	mm	mm	mm	mm	mm	mm
0·41	10·4140	10·4394	10·4648	10·4902	10·5156	10·5410	10·5664	10·5918	10·6172	10·6426
0·42	10·6680	10·6934	10·7188	10·7442	10·7696	10·7950	10·8204	10·8458	10·8712	10·8966
0·43	10·9220	10·9474	10·9728	10·9982	11·0236	11·0490	11·0744	11·0998	11·1252	11·1506
0·44	11·1760	11·2014	11·2268	11·2522	11·2776	11·3030	11·3284	11·3538	11·3792	11·4046
0·45	11·4300	11·4554	11·4808	11·5062	11·5316	11·5570	11·5824	11·6078	11·6332	11·6586
0·46	11·6840	11·7094	11·7348	11·7602	11·7856	11·8110	11·8364	11·8618	11·8872	11·9126
0·47	11·9380	11·9634	11·9888	12·0142	12·0396	12·0650	12·0904	12·1158	12·1412	12·1666
0·48	12·1920	12·2174	12·2428	12·2682	12·2936	12·3190	12·3444	12·3698	12·3952	12·4206
0·49	12·4460	12·4714	12·4968	12·5222	12·5476	12·5730	12·5984	12·6238	12·6492	12·6746
0·50	12·7000	12·7254	12·7508	12·7762	12·8016	12·8270	12·8524	12·8778	12·9032	12·9286
0·51	12·9540	12·9794	13·0048	13·0302	13·0556	13·0810	13·1064	13·1318	13·1572	13·1826
0·52	13·2080	13·2334	13·2588	13·2842	13·3096	13·3350	13·3604	13·3858	13·4112	13·4366
0·53	13·4620	13·4874	13·5128	13·5382	13·5636	13·5890	13·6144	13·6398	13·6652	13·6906
0·54	13·7160	13·7414	13·7668	13·7922	13·8176	13·8430	13·8684	13·8938	13·9192	13·9446
0·55	13·9700	13·9954	14·0208	14·0462	14·0716	14·0970	14·1224	14·1478	14·1732	14·1986
0·56	14·2240	14·2494	14·2748	14·3002	14·3256	14·3510	14·3764	14·4018	14·4272	14·4526
0·57	14·4780	14·5034	14·5288	14·5542	14·5796	14·6050	14·6304	14·6558	14·6812	14·7066
0·58	14·7320	14·7574	14·7828	14·8082	14·8336	14·8590	14·8844	14·9098	14·9352	14·9606
0·59	14·9860	15·0114	15·0368	15·0622	15·0876	15·1130	15·1384	15·1638	15·1892	15·2146
0·60	15·2400	15·2654	15·2908	15·3162	15·3416	15·3670	15·3924	15·4178	15·4432	15·4686

0·61	15·4940	15·5194	15·5448	15·5702	15·5956	15·6210	15·6464	15·6718	15·6972	15·7226
0·62	15·7480	15·7734	15·7988	15·8242	15·8496	15·8750	15·9004	15·9258	15·9512	15·9766
0·63	16·0020	16·0274	16·0528	16·0782	16·1036	16·1290	16·1544	16·1798	16·2052	16·2306
0·64	16·2560	16·2814	16·3068	16·3322	16·3576	16·3830	16·4084	16·4338	16·4592	16·4846
0·65	16·5100	16·5354	16·5608	16·5862	16·6116	16·6370	16·6624	16·6878	16·7132	16·7386
0·66	16·7640	16·7894	16·8148	16·8402	16·8656	16·8910	16·9164	16·9418	16·9672	16·9926
0·67	17·0180	17·0434	17·0688	17·0942	17·1196	17·1450	17·1704	17·1958	17·2212	17·2466
0·68	17·2720	17·2974	17·3228	17·3482	17·3736	17·3990	17·4244	17·4498	17·4752	17·5006
0·69	17·5260	17·5514	17·5768	17·6022	17·6276	17·6530	17·6784	17·7038	17·7292	17·7546
0·70	17·7800	17·8054	17·8308	17·8562	17·8816	17·9070	17·9324	17·9578	17·9832	18·0086
0·71	18·0340	18·0594	18·0848	18·1102	18·1356	18·1610	18·1864	18·2118	18·2372	18·2626
0·72	18·2880	18·3134	18·3388	18·3642	18·3896	18·4150	18·4404	18·4658	18·4912	18·5166
0·73	18·5420	18·5674	18·5928	18·6182	18·6436	18·6690	18·6944	18·7198	18·7452	18·7706
0·74	18·7960	18·8214	18·8468	18·8722	18·8976	18·9230	18·9484	18·9738	18·9992	19·0246
0·75	19·0500	19·0754	19·1008	19·1262	19·1516	19·1770	19·2024	19·2278	19·2532	19·2786
0·76	19·3040	19·3294	19·3548	19·3802	19·4056	19·4310	19·4564	19·4818	19·5072	19·5326
0·77	19·5580	19·5834	19·6088	19·6342	19·6596	19·6850	19·7104	19·7358	19·7612	19·7866
0·78	19·8120	19·8374	19·8628	19·8882	19·9136	19·9390	19·9644	19·9898	20·0152	20·0406
0·79	20·0660	20·0914	20·1168	20·1422	20·1676	20·1930	20·2184	20·2438	20·2692	20·2946
0·80	20·3200	20·3454	20·3708	20·3962	20·4216	20·4470	20·4724	20·4978	20·5232	20·5486
0·81	20·5740	20·5994	20·6248	20·6502	20·6756	20·7010	20·7264	20·7518	20·7772	20·8026
0·82	20·8280	20·8534	20·8788	20·9042	20·9296	20·9550	20·9804	21·0058	21·0312	21·0566
0·83	21·0820	21·1074	21·1328	21·1582	21·1836	21·2090	21·2344	21·2598	21·2852	21·3106
0·84	21·3360	21·3614	21·3868	21·4122	21·4376	21·4630	21·4884	21·5138	21·5392	21·5646
0·85	21·5900	21·6154	21·6408	21·6662	21·6916	21·7170	21·7424	21·7678	21·7932	21·8186

in	0·000	0·001	0·002	0·003	0·004	0·005	0·006	0·007	0·008	0·009
	mm	mm	mm	mm	mm	mm	mm	mm	mm	mm
0·86	21·8440	21·8694	21·8948	21·9202	21·9456	21·9710	21·9964	22·0218	22·0472	22·0726
0·87	22·0980	22·1234	22·1488	22·1742	22·1996	22·2250	22·2504	22·2758	22·3012	22·3266
0·88	22·3520	22·3774	22·4028	22·4282	22·4536	22·4790	22·5044	22·5298	22·5552	22·5806
0·89	22·6060	22·6314	22·6568	22·6822	22·7076	22·7330	22·7584	22·7838	22·8092	22·8346
0·90	22·8600	22·8854	22·9108	22·9362	22·9616	22·9870	23·0124	23·0378	23·0632	23·0886
0·91	23·1140	23·1394	23·1648	23·1902	23·2156	23·2410	23·2664	23·2918	23·3172	23·3426
0·92	23·3680	23·3934	23·4188	23·4442	23·4696	23·4950	23·5204	23·5458	23·5712	23·5966
0·93	23·6220	23·6474	23·6728	23·6982	23·7236	23·7490	23·7744	23·7998	23·8252	23·8506
0·94	23·8760	23·9014	23·9268	23·9522	23·9776	24·0030	24·0284	24·0538	24·0792	24·1046
0·95	24·1300	24·1554	24·1808	24·2062	24·2316	24·2570	24·2824	24·3078	24·3332	24·3586
0·96	24·3840	24·4094	24·4348	24·4602	24·4856	24·5110	24·5364	24·5618	24·5872	24·6126
0·97	24·6380	24·6634	24·6888	24·7142	24·7396	24·7650	24·7904	24·8158	24·8412	24·8666
0·98	24·8920	24·9174	24·9428	24·9682	24·9936	25·0190	25·0444	25·0698	25·0952	25·1206
0·99	25·1460	25·1714	25·1968	25·2222	25·2476	25·2730	25·2984	25·3238	25·3492	25·3746
1·00	25·4000									

Appendix 5

Miscellaneous technical data

CONVERSION TABLE

Pounds to Kilogrammes

lb	0·000	0·001	0·002	0·003	0·004	0·005	0·006	0·007	0·008	0·009
	Kg	Kg	Kg	Kg	Kg	Kg	Kg	Kg	Kg	Kg
0·00	0·0000	0·0005	0·0009	0·0014	0·0018	0·0023	0·0027	0·0032	0·0036	0·0041
0·01	0·0045	0·0050	0·0054	0·0059	0·0064	0·0068	0·0073	0·0077	0·0082	0·0086
0·02	0·0091	0·0095	0·0100	0·0104	0·0109	0·0113	0·0118	0·0122	0·0127	0·0132
0·03	0·0136	0·0141	0·0145	0·0150	0·0154	0·0159	0·0163	0·0168	0·0172	0·0177
0·04	0·0181	0·0186	0·0191	0·0195	0·0200	0·0204	0·0209	0·0213	0·0218	0·0222
0·05	0·0227	0·0231	0·0236	0·0240	0·0245	0·0249	0·0254	0·0258	0·0263	0·0268
0·06	0·0272	0·0277	0·0281	0·0286	0·0290	0·0295	0·0299	0·0304	0·0308	0·0313
0·07	0·0318	0·0322	0·0327	0·0331	0·0336	0·0340	0·0345	0·0349	0·0354	0·0358
0·08	0·0363	0·0367	0·0372	0·0376	0·0381	0·0386	0·0390	0·0395	0·0399	0·0404
0·09	0·0408	0·0413	0·0417	0·0422	0·0426	0·0431	0·0435	0·0440	0·0445	0·0449
0·10	0·0454	0·0458	0·0463	0·0467	0·0472	0·0476	0·0481	0·0485	0·0490	0·0494
0·11	0·0499	0·0503	0·0508	0·0513	0·0517	0·0522	0·0526	0·0531	0·0535	0·0540
0·12	0·0544	0·0549	0·0553	0·0558	0·0562	0·0567	0·0572	0·0576	0·0581	0·0585
0·13	0·0590	0·0594	0·0599	0·0603	0·0608	0·0612	0·0617	0·0621	0·0626	0·0630
0·14	0·0635	0·0640	0·0644	0·0649	0·0653	0·0658	0·0662	0·0667	0·0671	0·0676
0·15	0·0680	0·0685	0·0689	0·0694	0·0699	0·0703	0·0708	0·0712	0·0717	0·0721

lb	0·000	0·001	0·002	0·003	0·004	0·005	0·006	0·007	0·008	0·009
	Kg	Kg	Kg	Kg	Kg	Kg	Kg	Kg	Kg	Kg
0·16	0·0726	0·0730	0·0735	0·0739	0·0744	0·0748	0·0753	0·0757	0·0762	0·0767
0·17	0·0771	0·0776	0·0780	0·0785	0·0789	0·0794	0·0798	0·0803	0·0807	0·0812
0·18	0·0816	0·0821	0·0826	0·0830	0·0835	0·0839	0·0844	0·0848	0·0853	0·0857
0·19	0·0862	0·0866	0·0871	0·0875	0·0880	0·0885	0·0889	0·0894	0·0898	0·0903
0·20	0·0907	0·0912	0·0916	0·0921	0·0925	0·0930	0·0934	0·0939	0·0943	0·0948
0·21	0·0953	0·0957	0·0962	0·0966	0·0971	0·0975	0·0980	0·0984	0·0989	0·0993
0·22	0·0998	0·1002	0·1007	0·1012	0·1016	0·1021	0·1025	0·1030	0·1034	0·1039
0·23	0·1043	0·1048	0·1052	0·1057	0·1061	0·1066	0·1070	0·1075	0·1080	0·1084
0·24	0·1089	0·1093	0·1098	0·1102	0·1107	0·1111	0·1116	0·1120	0·1125	0·1129
0·25	0·1134	0·1139	0·1143	0·1148	0·1152	0·1157	0·1161	0·1166	0·1170	0·1175
0·26	0·1179	0·1184	0·1188	0·1193	0·1197	0·1202	0·1207	0·1211	0·1216	0·1220
0·27	0·1225	0·1229	0·1234	0·1238	0·1243	0·1247	0·1252	0·1256	0·1261	0·1266
0·28	0·1270	0·1275	0·1279	0·1284	0·1288	0·1293	0·1297	0·1302	0·1306	0·1311
0·29	0·1315	0·1320	0·1324	0·1329	0·1334	0·1338	0·1343	0·1347	0·1352	0·1356
0·30	0·1361	0·1365	0·1370	0·1374	0·1379	0·1383	0·1388	0·1393	0·1397	0·1402
0·31	0·1406	0·1411	0·1415	0·1420	0·1424	0·1429	0·1433	0·1438	0·1442	0·1447
0·32	0·1451	0·1456	0·1461	0·1465	0·1470	0·1474	0·1479	0·1483	0·1488	0·1492
0·33	0·1497	0·1501	0·1506	0·1510	0·1515	0·1520	0·1524	0·1529	0·1533	0·1538
0·34	0·1542	0·1547	0·1551	0·1556	0·1560	0·1565	0·1569	0·1574	0·1578	0·1583
0·35	0·1588	0·1592	0·1597	0·1601	0·1606	0·1610	0·1615	0·1619	0·1624	0·1628

0·36	0·1633	0·1637	0·1642	0·1647	0·1651	0·1656	0·1660	0·1665	0·1669	0·1674
0·37	0·1678	0·1683	0·1687	0·1692	0·1696	0·1701	0·1705	0·1710	0·1715	0·1719
0·38	0·1724	0·1728	0·1733	0·1737	0·1742	0·1746	0·1751	0·1755	0·1760	0·1764
0·39	0·1769	0·1774	0·1778	0·1783	0·1787	0·1792	0·1796	0·1801	0·1805	0·1810
0·40	0·1814	0·1819	0·1823	0·1828	0·1833	0·1837	0·1842	0·1846	0·1851	0·1855
0·41	0·1860	0·1864	0·1869	0·1873	0·1878	0·1882	0·1887	0·1891	0·1896	0·1901
0·42	0·1905	0·1910	0·1914	0·1919	0·1923	0·1928	0·1932	0·1937	0·1941	0·1946
0·43	0·1950	0·1955	0·1960	0·1964	0·1969	0·1973	0·1978	0·1982	0·1987	0·1991
0·44	0·1996	0·2000	0·2005	0·2009	0·2014	0·2018	0·2023	0·2028	0·2032	0·2037
0·45	0·2041	0·2046	0·2050	0·2055	0·2059	0·2064	0·2068	0·2073	0·2077	0·2082
0·46	0·2087	0·2091	0·2096	0·2100	0·2105	0·2109	0·2114	0·2118	0·2123	0·2127
0·47	0·2132	0·2136	0·2141	0·2145	0·2150	0·2155	0·2159	0·2164	0·2168	0·2173
0·48	0·2177	0·2182	0·2186	0·2191	0·2195	0·2200	0·2204	0·2209	0·2214	0·2218
0·49	0·2223	0·2227	0·2232	0·2236	0·2241	0·2245	0·2250	0·2254	0·2259	0·2263
0·50	0·2268	0·2272	0·2277	0·2282	0·2286	0·2291	0·2295	0·2300	0·2304	0·2309
0·51	0·2313	0·2318	0·2322	0·2327	0·2331	0·2336	0·2341	0·2345	0·2350	0·2354
0·52	0·2359	0·2363	0·2368	0·2372	0·2377	0·2381	0·2386	0·2390	0·2395	0·2399
0·53	0·2404	0·2409	0·2413	0·2418	0·2422	0·2427	0·2431	0·2436	0·2440	0·2445
0·54	0·2449	0·2454	0·2458	0·2463	0·2468	0·2472	0·2477	0·2481	0·2486	0·2490
0·55	0·2495	0·2499	0·2504	0·2508	0·2513	0·2517	0·2522	0·2526	0·2531	0·2536
0·56	0·2540	0·2545	0·2549	0·2554	0·2558	0·2563	0·2567	0·2572	0·2576	0·2581
0·57	0·2585	0·2590	0·2595	0·2599	0·2604	0·2608	0·2613	0·2617	0·2622	0·2626
0·58	0·2631	0·2635	0·2640	0·2644	0·2649	0·2654	0·2658	0·2663	0·2667	0·2672
0·59	0·2676	0·2681	0·2685	0·2690	0·2694	0·2699	0·2703	0·2708	0·2712	0·2717
0·60	0·2722	0·2726	0·2731	0·2735	0·2740	0·2744	0·2749	0·2753	0·2758	0·2762

lb	0·000	0·001	0·002	0·003	0·004	0·005	0·006	0·007	0·008	0·009
	Kg	Kg	Kg	Kg	Kg	Kg	Kg	Kg	Kg	Kg
0·61	0·2767	0·2771	0·2776	0·2781	0·2785	0·2790	0·2794	0·2799	0·2803	0·2808
0·62	0·2812	0·2817	0·2821	0·2826	0·2830	0·2835	0·2839	0·2844	0·2849	0·2853
0·63	0·2858	0·2862	0·2867	0·2871	0·2876	0·2880	0·2885	0·2889	0·2894	0·2898
0·64	0·2903	0·2908	0·2912	0·2917	0·2921	0·2926	0·2930	0·2935	0·2939	0·2944
0·65	0·2948	0·2953	0·2957	0·2962	0·2966	0·2971	0·2976	0·2980	0·2985	0·2989
0·66	0·2994	0·2998	0·3003	0·3007	0·3012	0·3016	0·3021	0·3025	0·3030	0·3035
0·67	0·3039	0·3044	0·3048	0·3053	0·3057	0·3062	0·3066	0·3071	0·3075	0·3080
0·68	0·3084	0·3089	0·3093	0·3098	0·3103	0·3107	0·3112	0·3116	0·3121	0·3125
0·69	0·3130	0·3134	0·3139	0·3143	0·3148	0·3152	0·3157	0·3162	0·3166	0·3171
0·70	0·3175	0·3180	0·3184	0·3189	0·3193	0·3198	0·3202	0·3207	0·3211	0·3216
0·71	0·3220	0·3225	0·3230	0·3234	0·3239	0·3243	0·3248	0·3252	0·3257	0·3261
0·72	0·3266	0·3270	0·3275	0·3279	0·3284	0·3289	0·3293	0·3298	0·3302	0·3307
0·73	0·3311	0·3316	0·3320	0·3325	0·3329	0·3334	0·3338	0·3343	0·3347	0·3352
0·74	0·3357	0·3361	0·3366	0·3370	0·3375	0·3379	0·3384	0·3388	0·3393	0·3397
0·75	0·3402	0·3406	0·3411	0·3416	0·3420	0·3425	0·3429	0·3434	0·3438	0·3443
0·76	0·3447	0·3452	0·3456	0·3461	0·3465	0·3470	0·3475	0·3479	0·3484	0·3488
0·77	0·3493	0·3497	0·3502	0·3506	0·3511	0·3515	0·3520	0·3524	0·3529	0·3533
0·78	0·3538	0·3543	0·3547	0·3552	0·3556	0·3561	0·3565	0·3570	0·3574	0·3579
0·79	0·3583	0·3588	0·3592	0·3597	0·3602	0·3606	0·3611	0·3615	0·3620	0·3624
0·80	0·3629	0·3633	0·3638	0·3642	0·3647	0·3651	0·3656	0·3660	0·3665	0·3670

0·81	0·3674	0·3679	0·3683	0·3688	0·3692	0·3697	0·3701	0·3706	0·3710	0·3715
0·82	0·3719	0·3724	0·3729	0·3733	0·3738	0·3742	0·3747	0·3751	0·3756	0·3760
0·83	0·3765	0·3769	0·3774	0·3778	0·3783	0·3787	0·3792	0·3797	0·3801	0·3806
0·84	0·3810	0·3815	0·3819	0·3824	0·3828	0·3833	0·3837	0·3842	0·3846	0·3851
0·85	0·3856	0·3860	0·3865	0·3869	0·3874	0·3878	0·3883	0·3887	0·3892	0·3896
0·86	0·3901	0·3905	0·3910	0·3914	0·3919	0·3924	0·3928	0·3933	0·3937	0·3942
0·87	0·3946	0·3951	0·3955	0·3960	0·3964	0·3969	0·3973	0·3978	0·3983	0·3987
0·88	0·3992	0·3996	0·4001	0·4005	0·4010	0·4014	0·4019	0·4023	0·4028	0·4032
0·89	0·4037	0·4041	0·4046	0·4051	0·4055	0·4060	0·4064	0·4069	0·4073	0·4078
0·90	0·4082	0·4087	0·4091	0·4096	0·4100	0·4105	0·4110	0·4114	0·4119	0·4123
0·91	0·4128	0·4132	0·4137	0·4141	0·4146	0·4150	0·4155	0·4159	0·4164	0·4168
0·92	0·4173	0·4178	0·4182	0·4187	0·4191	0·4196	0·4200	0·4205	0·4209	0·4214
0·93	0·4218	0·4223	0·4227	0·4232	0·4237	0·4241	0·4246	0·4250	0·4255	0·4259
0·94	0·4264	0·4268	0·4273	0·4277	0·4282	0·4286	0·4291	0·4295	0·4300	0·4305
0·95	0·4309	0·4314	0·4318	0·4323	0·4327	0·4332	0·4336	0·4341	0·4345	0·4350
0·96	0·4354	0·4359	0·4364	0·4368	0·4373	0·4377	0·4382	0·4386	0·4391	0·4395
0·97	0·4400	0·4404	0·4409	0·4413	0·4418	0·4423	0·4427	0·4432	0·4436	0·4441
0·98	0·4445	0·4450	0·4454	0·4459	0·4463	0·4468	0·4472	0·4477	0·4481	0·4486
0·99	0·4491	0·4495	0·4500	0·4504	0·4509	0·4513	0·4518	0·4522	0·4527	0·4531
1·00	0·4536									

Appendix 6

CONVERSION TABLE

Pounds per Foot to Kilogrammes per Metre

Miscellaneous technical data

lb/ft	0·000	0·001	0·002	0·003	0·004	0·005	0·006	0·007	0·008	0·009
	Kg/m	Kg/m	Kg/m	Kg/m	Kg/m	Kg/m	Kg/m	Kg/m	Kg/m	Kg/m
0·00	0·0000	0·0015	0·0030	0·0045	0·0060	0·0074	0·0089	0·0104	0·0119	0·0134
0·01	0·0149	0·0164	0·0179	0·0193	0·0208	0·0223	0·0238	0·0253	0·0268	0·0283
0·02	0·0298	0·0313	0·0327	0·0342	0·0357	0·0372	0·0387	0·0402	0·0417	0·0432
0·03	0·0446	0·0461	0·0476	0·0491	0·0506	0·0521	0·0536	0·0551	0·0566	0·0580
0·04	0·0595	0·0610	0·0625	0·0640	0·0655	0·0670	0·0685	0·0699	0·0714	0·0729
0·05	0·0744	0·0759	0·0774	0·0789	0·0804	0·0819	0·0833	0·0848	0·0863	0·0878
0·06	0·0893	0·0908	0·0923	0·0938	0·0952	0·0967	0·0982	0·0997	0·1012	0·1027
0·07	0·1042	0·1057	0·1072	0·1086	0·1101	0·1116	0·1131	0·1146	0·1161	0·1176
0·08	0·1191	0·1205	0·1220	0·1235	0·1250	0·1265	0·1280	0·1295	0·1310	0·1325
0·09	0·1339	0·1354	0·1369	0·1384	0·1399	0·1414	0·1429	0·1444	0·1458	0·1473
0·10	0·1488	0·1503	0·1518	0·1533	0·1548	0·1563	0·1577	0·1592	0·1607	0·1622
0·11	0·1637	0·1652	0·1667	0·1682	0·1697	0·1711	0·1726	0·1741	0·1756	0·1771
0·12	0·1786	0·1801	0·1816	0·1830	0·1845	0·1860	0·1875	0·1890	0·1905	0·1920
0·13	0·1935	0·1949	0·1964	0·1979	0·1994	0·2009	0·2024	0·2039	0·2054	0·2069
0·14	0·2083	0·2098	0·2113	0·2128	0·2143	0·2158	0·2173	0·2188	0·2202	0·2217
0·15	0·2232	0·2247	0·2262	0·2277	0·2292	0·2307	0·2322	0·2336	0·2351	0·2366

0·16	0·2381	0·2396	0·2411	0·2426	0·2441	0·2455	0·2470	0·2485	0·2500	0·2515
0·17	0·2530	0·2545	0·2560	0·2575	0·2589	0·2604	0·2619	0·2634	0·2649	0·2664
0·18	0·2679	0·2694	0·2708	0·2723	0·2738	0·2753	0·2768	0·2783	0·2798	0·2813
0·19	0·2828	0·2842	0·2857	0·2872	0·2887	0·2902	0·2917	0·2932	0·2947	0·2961
0·20	0·2976	0·2991	0·3006	0·3021	0·3036	0·3051	0.3066	0·3080	0·3095	0·3110
0·21	0·3125	0·3140	0·3155	0·3170	0·3185	0·3200	0·3214	0·3229	0·3244	0·3259
0·22	0·3274	0·3289	0·3304	0·3319	0·3333	0·3348	0·3363	0·3378	0·3393	0·3408
0·23	0·3423	0·3438	0·3453	0·3467	0·3482	0·3497	0·3512	0·3527	0·3542	0·3557
0·24	0·3572	0·3586	0·3601	0·3616	0·3631	0·3646	0·3661	0·3676	0·3691	0·3706
0·25	0·3720	0·3735	0·3750	0·3765	0·3780	0·3795	0·3810	0·3825	0·3839	0·3854
0·26	0·3869	0·3884	0·3899	0·3914	0·3929	0·3944	0·3959	0·3973	0·3988	0·4003
0·27	0·4018	0·4033	0·4048	0·4063	0·4078	0·4092	0·4107	0·4122	0·4137	0·4152
0·28	0·4167	0·4182	0·4197	0·4211	0·4226	0·4241	0·4256	0·4271	0·4286	0·4301
0·29	0·4316	0·4331	0·4345	0·4360	0·4375	0·4390	0·4405	0·4420	0·4435	0·4450
0·30	0·4464	0·4479	0·4494	0·4509	0·4524	0·4539	0·4554	0·4569	0·4584	0·4598
0·31	0·4613	0·4628	0·4643	0·4658	0·4673	0·4688	0·4703	0·4717	0·4732	0·4747
0·32	0·4762	0·4777	0·4792	0·4807	0·4822	0·4837	0·4851	0·4866	0·4881	0·4896
0·33	0·4911	0·4926	0·4941	0·4956	0·4970	0·4985	0·5000	0·5015	0·5030	0·5045
0·34	0·5060	0·5075	0·5090	0·5104	0.5119	0·5134	0·5149	0·5164	0·5179	0·5194
0·35	0·5209	0·5223	0·5238	0·5253	0·5268	0·5283	0·5298	0·5313	0·5328	0·5342
0·36	0·5357	0·5372	0·5387	0·5402	0·5417	0·5432	0·5447	0·5462	0·5476	0·5491
0·37	0·5506	0·5521	0·5536	0·5551	0·5566	0·5581	0·5595	0·5610	0·5625	0·5640
0·38	0·5655	0·5670	0·5685	0·5700	0·5715	0·5729	0·5744	0·5759	0·5774	0·5789
0·39	0·5804	0·5819	0·5834	0·5848	0·5863	0·5878	0·5893	0·5908	0·5923	0·5938
0·40	0·5953	0·5968	0·5982	0·5997	0·6012	0·6027	0·6042	0·6057	0·6072	0·6087

lb/ft	0·000	0·001	0·002	0·003	0·004	0·005	0·006	0·007	0·008	0·009
	Kg/m	Kg/m	Kg/m	Kg/m	Kg/m	Kg/m	Kg/m	Kg/m	Kg/m	Kg/m
0·41	0·6101	0·6116	0·6131	0·6146	0·6161	0·6176	0·6191	0·6206	0·6221	0·6235
0·42	0·6250	0·6265	0·6280	0·6295	0·6310	0·6325	0·6340	0·6354	0·6369	0·6384
0·43	0·6399	0·6414	0·6429	0·6444	0·6459	0·6473	0·6488	0·6503	0·6518	0·6533
0·44	0·6548	0·6563	0·6578	0·6593	0·6607	0·6622	0·6637	0·6652	0·6667	0·6682
0·45	0·6697	0·6712	0·6726	0·6741	0·6756	0·6771	0·6786	0·6801	0·6816	0·6831
0·46	0·6846	0·6860	0·6875	0·6890	0·6905	0·6920	0·6935	0·6950	0·6965	0·6979
0·47	0·6994	0·7009	0·7024	0·7039	0·7054	0·7069	0·7084	0·7099	0·7113	0·7128
0·48	0·7143	0·7158	0·7173	0·7188	0·7203	0·7218	0·7232	0·7247	0·7262	0·7277
0·49	0·7292	0·7307	0·7322	0·7337	0·7352	0·7366	0·7381	0·7396	0·7411	0·7426
0·50	0·7441	0·7456	0·7471	0·7485	0·7500	0·7515	0·7530	0·7545	0·7560	0·7575
0·51	0·7590	0·7604	0·7619	0·7634	0·7649	0·7664	0·7679	0·7694	0·7709	0·7724
0·52	0·7738	0·7753	0·7768	0·7783	0·7798	0·7813	0·7828	0·7843	0·7857	0·7872
0·53	0·7887	0·7902	0·7917	0·7932	0·7947	0·7961	0·7977	0·7991	0·8006	0·8021
0·54	0·8036	0·8051	0·8066	0·8081	0·8096	0·8110	0·8125	0·8140	0·8155	0·8170
0·55	0·8185	0·8200	0·8215	0·8230	0·8244	0·8259	0·8274	0·8289	0·8304	0·8319
0·56	0·8334	0·8349	0·8363	0·8378	0·8393	0·8408	0·8423	0·8438	0·8453	0·8468
0·57	0·8483	0·8497	0·8512	0·8527	0·8542	0·8557	0·8572	0·8587	0·8602	0·8616
0·58	0·8631	0·8646	0·8661	0·8676	0·8691	0·8706	0·8721	0·8735	0·8750	0·8765
0·59	0·8780	0·8795	0·8810	0·8825	0·8840	0·8855	0·8869	0·8884	0·8899	0·8914
0·60	0·8929	0·8944	0·8959	0·8974	0·8988	0·9003	0·9018	0·9033	0·9048	0·9063

0·61	0·9078	0·9093	0·9108	0·9122	0·9137	0·9152	0·9167	0·9182	0·9197	9·9212
0·62	0·9227	0·9241	0·9256	0·9271	0·9286	0·9301	0·9316	0·9331	0·9346	0·9361
0·63	0·9375	0·9390	0·9405	0·9420	0·9435	0·9450	0·9465	0·9480	0·9494	0·9509
0·64	0·9524	0·9539	0·9554	0·9569	0·9584	0·9599	0·9614	0·9628	0·9643	0·9658
0·65	0·9673	0·9688	0·9703	0·9718	0·9733	0·9747	0·9762	0·9777	0·9792	0·9807
0·66	0·9822	0·9837	0·9852	0·9867	0·9881	0·9896	0·9911	0·9926	0·9941	0·9956
0·67	0·9971	0·9986	1·0000	1·0015	1·0030	1·0045	1·0060	1·0075	1·0090	1·0105
0·68	1·0119	1·0134	1·0149	1·0164	1·0179	1·0194	1·0209	1·0224	1·0239	1·0253
0·69	1·0268	1·0283	1·0298	1·0313	1·0328	1·0343	1·0358	1·0372	1·0387	1·0402
0·70	1·0417	1·0432	1·0447	1·0462	1·0477	1·0492	1·0506	1·0521	1·0536	1·0551
0·71	1·0566	1·0581	1·0596	1·0611	1·0625	1·0640	1·0655	1·0670	1·0685	1·0700
0·72	1·0715	1·0730	1·0745	1·0759	1·0774	1·0789	1·0804	1·0819	1·0834	1·0847
0·73	1·0864	1·0878	1·0893	1·0908	1·0923	1·0938	1·0953	1·0968	1·0983	1·0998
0·74	1·1012	1·1027	1·1042	1·1057	1·1072	1·1087	1·1102	1·1117	1·1131	1·1146
0·75	1·1161	1·1176	1·1190	1·1206	1·1221	1·1236	1·1250	1·1265	1·1280	1·1295
0·76	1·1310	1·1325	1·1340	1·1355	1·1370	1·1384	1·1399	1·1414	1·1429	1·1444
0·77	1·1459	1·1474	1·1489	1·1503	1·1518	1·1533	1·1548	1·1563	1·1578	1·1593
0·78	1·1608	1·1623	1·1637	1·1652	1·1667	1·1682	1·1697	1·1712	1·1727	1·1742
0·79	1·1756	1·1771	1·1786	1·1801	1·1816	1·1831	1·1846	1·1861	1·1876	1·1890
0·80	1·1905	1·1920	1·1935	1·1950	1·1965	1·1980	1·1995	1·2009	1·2024	1·2039
0·81	1·2054	1·2069	1·2084	1·2099	1·2114	1·2129	1·2143	1·2158	1·2173	1·2188
0·82	1·2203	1·2218	1·2233	1·2248	1·2262	1·2277	1·2292	1·2307	1·2322	1·2337
0·83	1·2352	1·2367	1·2381	1·2396	1·2411	1·2426	1·2441	1·2456	1·2471	1·2486
0·84	1·2500	1·2515	1·2530	1·2545	1·2560	1·2575	1·2590	1·2605	1·2620	1·2634
0·85	1·2649	1·2664	1·2679	1·2694	1·2709	1·2724	1·2739	1·2754	1·2768	1·2783

lb/ft	0·000	0·001	0·002	0·003	0·004	0·005	0·006	0·007	0·008	0·009
	Kg/m	Kg/m	Kg/m	Kg/m	Kg/m	Kg/m	Kg/m	Kg/m	Kg/m	Kg/m
0·86	1·2798	1·2813	1·2828	1·2843	1·2858	1·2873	1·2887	1·2902	1·2917	1·2932
0·87	1·2947	1·2962	1·2977	1·2992	1·3007	1·3021	1·3036	1·3051	1·3066	1·3081
0·88	1·3096	1·3111	1·3126	1·3140	1·3155	1·3170	1·3185	1·3200	1·3215	1·3230
0·89	1·3245	1·3260	1·3274	1·3289	1·3304	1·3319	1·3334	1·3349	1·3364	1·3379
0·90	1·3393	1·3408	1·3423	1·3438	1·3453	1·3468	1·3483	1·3498	1·3513	1·3527
0·91	1·3542	1·3557	1·3572	1·3587	1·3602	1·3617	1·3632	1·3646	1·3661	1·3676
0·92	1·3691	1·3706	1·3721	1·3736	1·3751	1·3765	1·3780	1·3795	1·3810	1·3825
0·93	1·3840	1·3855	1·3870	1·3885	1·3899	1·3914	1·3929	1·3944	1·3959	1·3974
0·94	1·3989	1·4004	1·4018	1·4033	1·4048	1·4063	1·4078	1·4093	1·4108	1·4123
0·95	1·4138	1·4152	1·4167	1·4182	1·4197	1·4212	1·4227	1·4242	1·4257	1·4271
0·96	1·4286	1·4301	1·4316	1·4331	1·4346	1·4361	1·4376	1·4391	1·4405	1·4420
0·97	1·4435	1·4450	1·4465	1·4480	1·4495	1·4510	1·4524	1·4539	1·4554	1·4569
0·98	1·4584	1·4599	1·4614	1·4629	1·4644	1·4658	1·4673	1·4688	1·4703	1·4718
0·99	1·4733	1·4748	1·4763	1·4777	1·4792	1·4807	1·4822	1·4837	1·4852	1·4867
1·00	1·4882									

Index

Plates appear in italics